Scientific Protocols for

FORENSIC EXAMINATION OF CLOTHING

Scientific Protocols for Forensic Examination of Clothing
by Jane Moira Taupin and Chesterene Cwiklik

Principles and Practices of Criminalistics: The Profession of Forensic Science
by Keith Inman and Norah Rudin

Ethics in Forensic Science: Professional Standards for the Practice of Criminalistics
by Peter D. Barnett

Scientific Protocols for Fire Investigation
by John J. Lentini

Scientific Protocols for FORENSIC EXAMINATION OF CLOTHING

Jane Moira Taupin
Chesterene Cwiklik

CRC Press is an imprint of the
Taylor & Francis Group, an **informa** business

CRC Press
Taylor & Francis Group
6000 Broken Sound Parkway NW, Suite 300
Boca Raton, FL 33487-2742

Printed in the United States of America on acid-free paper
10 9 8 7 6 5 4 3 2 1

International Standard Book Number: 978-1-4200-6821-4 (Hardback)

Library of Congress Cataloging-in-Publication Data

Taupin, Jane Moira.
Scientific protocols for forensic examination of clothing / Jane Moira Taupin, Chesterene Cwiklik.
p. cm. -- (Protocols in forensic science series)
Includes bibliographical references and index.
ISBN 978-1-4200-6821-4 (hardcover : alk. paper)
1. Forensic sciences. 2. Evidence, Criminal. 3. Clothing and dress. 4. Criminal investigation. I. Cwiklik, Chesterene. II. Title.

HV8077.5.C6T38 2011
363.25'62--dc22 2010042903

Visit the Taylor & Francis Web site at
http://www.taylorandfrancis.com

and the CRC Press Web site at
http://www.crcpress.com

Table of Contents

Preface

This text will describe and discuss the forensic examination of clothing, primarily clothing exhibits in criminal and civil cases. Clothing and other textiles are part of everyday life; so it is not surprising that when a crime or other incident takes place, clothing items are present and often directly involved. Items of clothing are thus one of the most common types of exhibit examined. Clothing can provide valuable information in cases of violent crimes, such as homicide or rape, and in burglary, robbery, arson, vehicular accidents, and other crimes and infractions. Clothing items often contain crucial evidence. Moreover, examination of clothing from a crime may elicit its "story," much like examining a crime scene helps to reconstruct the crime. In some cases, the garment itself may be considered a crime scene.

No comprehensive text on forensic examination of clothing exists. The subject has traditionally been presented as a chapter in general forensic texts or been discussed as a source of samples when describing specialized forensic techniques. This text focuses on the clothing itself, including damage to the clothing and information from stains and deposits encountered on it.

Because clothing submitted as potential evidence has most often been worn on the body, special features must be considered. The garment may have traveled from one scene to another, may not have been stationary during the commission of the crime, or may even have been worn by someone else prior to, or subsequent to, the crime or other incident under investigation. There may be damage associated with deposits and deposits associated with particular actions and body movements.

We believe that the subject of clothing examination deserves a comprehensive treatment. As DNA testing technology becomes increasingly specific to individuals, and as increasingly smaller amounts of DNA can be coaxed to yield results, the sampling of evidence draws our attention. By examining damage and deposits and evaluating what actions may have produced them, the examiner can ensure that the samples collected for testing have the potential to address the questions in a case. This defines the potential for a significant test result, giving the examiner a tool to control for error and to make defensible testing decisions that can withstand scrutiny. Preliminary examinations and interpretations form the basis for subsequent testing. Rigorous sampling decisions ensure that subsequent testing is relevant and useful. In addition, the information obtained from clothing examination may provide answers about the circumstances of a case. We hope this book provides the examiner with some tools for these tasks.

Every criminalist or forensic scientist who analyzes samples from clothing items should know how to examine clothing to discover the relevant evidence and understand how that evidence relates to crucial legal questions. Other personnel who collect evidence from clothing items, including forensic pathologists or police evidence technicians, should approach the task with similar understanding. We will describe crucial factors to consider when analyzing a clothing item.

Some forensic laboratories assign cases to forensic scientists or caseworkers with a particular specialty, because the evidence of initial interest is in that field. For example, a biologist may receive clothing from a rape case, or a firearms examiner clothing from a shooting. That caseworker is then responsible for recognizing and preserving all potentially relevant evidence, not only the evidence in that specialty. Other laboratories designate a generalist to perform a complete clothing examination that includes sample collection, or technicians may collect samples from clothing and submit them to the particular forensic specialists. In addition, the police, other investigators, or evidence technicians may submit samples from clothing to a specialist. We hope that all examiners of clothing will find the text useful and thought provoking.

Forensic scientists are not expected to be experts in any or all specialized forensic techniques. Forensic specialties, such as DNA profiling or fire scene reconstruction, may require considerable study and qualification before one can be considered an expert. Unless qualified in more than one specialty, a forensic biologist who is given a T-shirt as evidence from a shooting would not be expected to analyze gunshot residue obtained from that T-shirt, and in some laboratories would not be expected to perform DNA profiling on blood from it. However, he or she *should* be expected to *recognize* the presence of gunshot residue and make the necessary accommodations for its analysis. The initial examiner should also be able to integrate the results of subsequent testing with the data from the initial clothing examination without exceeding the bounds of his or her own expertise.

The work of the Innocence Project at Cardozo Law School in New York, through its use of DNA technology, has led to the courts overturning or dismissing the convictions of more than 200 prisoners, many of whom were on death row (Scheck et al., 2000; Innocence Project, 2008). Some of these convictions were attributed to incomplete or erroneous interpretation of forensic evidence, including clothing items. In this imperfect world, it is the concern of many a good scientist that he or she may have overlooked evidence that may be significant or may have misinterpreted a test result. This has stimulated our work on a systematic, data-based approach that acknowledges the importance of and confers rigor on non-numerical observations.

We are pleased that this text is part of the Protocols in Forensic Science Series, edited by Keith Inman and Norah Rudin. Their flagship book, *Principles and Practice of Criminalistics* (Inman and Rudin, 2001), eloquently expressed the basic principles of forensic examination. A good scientific protocol encompasses an approach to examination that is grounded in sound scientific practice and encompasses the scientific basis, advantages, and limitations of each technique or step in the process. We hope that this text will help the examiner who is trying to decide how to approach a clothing examination, that it will be a resource for the examiner who would like more information on a specific topic, and that it will be a useful reference for the laboratory quality assurance officer.

It is not necessary to read and digest the whole book to gain information as to why clothing is examined and what information can be obtained in the context of a particular crime. Thus, an attorney or investigator may gain information that may help the questioning of an expert witness or may assist in deciphering a forensic report or statement by referring to the relevant forensic discipline and/or chapters as outlined in the table of contents or index. We hope the text will be useful to the police detective, investigator, attorney, archaeologist, or curator who needs to understand the types of information that can be obtained from clothing examination.

In closing, this book provides a comprehensive, integrated, interdisciplinary approach to the examination of clothing that can be used as a ready reference when examining a clothing exhibit.

References

Inman, K. and Rudin, N., *Principles and Practice of Criminalistics: The Profession of Forensic Science*, CRC Press, Boca Raton, FL, 2001.

Innocence Project, http://www.innocenceproject.org/ for more recent information and statistics; accessed May 2008.

Scheck B., Neufeld, P., and Dwyer, J., *Actual Innocence*, Doubleday, New York, 2000.

Acknowledgments

The authors give their appreciation to Norah Rudin and Keith Inman, the editors of this series, for their willingness to take on this topic. Thanks again to Norah and Keith for their time spent reviewing the work. We also thank Becky Masterman, our editor at Taylor & Francis, for her unending patience and support.

Jane Moira Taupin would like to thank LGC Forensics England for permission to use case studies in this book. In particular, she would like to thank the director, Dr. Angela Gallop, for her scholarly and innovative approach to casework, which has resulted in a supportive environment for the forensic scientists in that company. She would also like to thank LGC Forensics photographer Richard Thomas and the many biologists who contributed case studies to this book, including Deb Hopwood, Dr. Lisa Edwards, Tom Jalowiecki, Luan Lunt, Claire Stangoe, and Pauline Stevens.

Thanks also to Dr. Roland van Oorschot of the Victoria Police Forensic Services Department for his support over many years in the publication of case studies by a practicing forensic scientist, and Dr. Tony Raymond, previously of the same laboratory, for inspiring her with his simulation approach to clothing damage casework. She has appreciated the personal kindness, friendship, and sharing of valuable knowledge on fiber evidence transfer by the late Mike Grieve.

And finally, she thanks Peter Barnett for the idea that led to this book.

Chesterene Cwiklik extends her appreciation first to Mary Jarrett-Jackson, formerly the supervisor of the serology and trace evidence unit of the Detroit Police Department Crime Laboratory, who taught her how to look at a case and how to perform a clothing examination. She thanks George Ishii, who as the laboratory director of the Washington State Patrol Crime Laboratory in Seattle provided a clear focus on the crime laboratory, serving a criminal justice system, including police, prosecution, and defense. She thanks Kay Sweeney, head of the criminalistics section and later director of the same laboratory for fostering a multidisciplinary approach to casework and ensuring the cross-training of scientists in disciplines outside their own. Mr. Sweeney is now with KMS Forensics, Inc., and continues to be a valued colleague. Mrs. Jarrett-Jackson, Mr. Ishii, and Mr. Sweeney, through their insights and questions, fostered a spirit of inquiry, and by their management decisions, encouraged thorough casework and supported research. She is grateful to Dr. Walter C. McCrone for encouraging her interest in microscopy, and thanks Wheeling College chemistry professors Chester A. Giza, PhD, and Charles J. Loner, PhD, for insisting on rigorous scientific thinking.

She also thanks the numerous colleagues who have provided ideas and discussion, and thanks Richard E. Bisbing, Carol Murren, Terry M. Franklin, William R. Gresham, PhD, Helen R. Griffin, Kay M. Sweeney, and Skip Palenik, PhD, for their thoughtful comments on portions of this book. Special thanks to colleagues who have collaborated in clothing examination, including John A. Brown, PhD, George K. Chan, Kerstin M. Gleim, Michael J. Grubb, Jennifer Iem, Vanora Kean, PhD, Cindi B. Jay, Mary Jarrett-Jackson, George E. Johnston, Carol Murren, Linda D. McGarvey, Lynn D. McIntyre, and Kay M. Sweeney. She also thanks Mary Jarrett-Jackson, Lynn McIntyre, George Chan, and Kerstin Gleim for case examples used in this book.

She thanks the staff and former staff of the California Institute of Criminalistics, especially Theresa F. Spear and Cecilia von Beroldingen, for their support in developing the clothing examination class, and Neda Khoshkebari, for support in implementing it. She thanks her colleagues and co-workers who were willing to try out the ideas in this book, especially forensic scientists Jennifer Iem and Linda D. McGarvey and forensic science assistant Melanie (Kubi) Thomas. Finally, she thanks her students, who applied and refined the terminology for stains and deposits.

About the Authors

Jane Moira Taupin obtained a bachelor of science (honors) degree from the University of Melbourne in Australia. Upon graduating, she accepted research positions at University of Melbourne research facilities, first in antibody production at the Howard Florey Institute and in cancer research at the Austin Hospital. She joined the Australian Federal Police as a constable and then stage 1 detective and worked in diverse areas, including drug surveillance and government fraud. During this time, she was transferred temporarily to work at the only atomic energy facility in the country (Lucas Heights), using neutron activation analysis on a number of criminal cases. She left to join the Victoria Police Forensic Services Centre, initially working in the blood alcohol section, where she performed blood alcohol analyses. She transferred to the biology section, where she reported on a wide variety of cases involving biological evidence in major crime; this included attendance at scenes of crime and presenting expert evidence in courts of law. Concurrently, she obtained a postgraduate diploma in criminology and a master of arts in criminology, both from the University of Melbourne. She joined Forensic Alliance in England, where she performed similar work. She was employed at LGC Forensics in England, where she was a lead scientist. She has now returned to Australia where she is a forensic consultant and auditor with MRS Limited (www.mrslimited.com).

As a result of her presentation of a case study on hair and fiber transfer to the Australian and New Zealand Forensic Science Society annual meeting in 1994 in Auckland, she published a paper in the *Journal of Forensic Sciences,* which sparked her interest in the publication of case studies for the working forensic scientist. For her work on clothing damage analysis, she won a Young Investigators Award from the International Association of Forensic Sciences to attend their meeting in Tokyo in 1996. The following year, in recognition of her work on clothing damage and hairs, she was awarded an Australian Government Michael Duffy Travel Fellowship to attend international laboratories as well as the American Academy of Forensic Sciences meeting in New York. She visited the John Jay School of Criminal Justice, the FBI Laboratory and Academy, the Metropolitan Police Laboratory in London, and the Bundeskriminilat in Wiesbaden. She was invited to be on the inaugural committee of SWGHAIR (Scientific Working Group on Hair) under the auspices of the FBI, where she met her co-author Chesterene Cwiklik. Her main forensic interests are clothing damage, fiber and hair transfer, and blood pattern analysis at crime scenes.

Chesterene Cwiklik has been a forensic scientist in private practice since 1990 with Cwiklik & Associates, a laboratory specializing in trace evidence and small particle analysis, general criminalistics including incident reconstruction, and forensic consulting. Serving on the board and faculty of the Pacific Coast Forensic Science Institute since its inception in 1998, she is dedicated to teaching and research in forensic science. She has testified as an expert witness in numerous complex and high-profile cases, both criminal and civil, and has worked with the prosecution, plaintiff, and defense. She earned a bachelor of science in chemistry from Wheeling College in West Virginia (now Wheeling Jesuit University) and did postgraduate work in organic chemistry at Wayne State University in Detroit. She began her career with the Detroit Police Department Crime Laboratory, beginning in the chemistry section analyzing controlled substances and developing latent prints, then transferring to the serology and trace evidence unit. Her interest in clothing examination was cemented while she was in the supervised casework portion of training and working on a narrow portion of a complex case with unknown suspects and an unknown scene. In the initial discussion, her supervisor, Mary Jarrett-Jackson, looked at a deposit and said, "Bet you a quarter that's a bean soup splash." She then sent the scientists in the unit home to cook a variety of beans. Together with trace evidence and bakery sheets, this eventually led police to the scene of two apparently separate murders.

Ms. Cwiklik later joined the Washington State Patrol Crime Laboratory in Seattle, where she set up and developed the trace evidence analysis program and was the head of the microanalysis section for 14 years. That section contained a fledgling program of methods development for analysis of incendiary materials from fire scenes. She collaborated closely with her supervisor and her peers in other sections, coordinating work, developing priority systems, evaluating equipment and recommending purchases, and implementing cross-training in different areas of specialty. She served on the Scientific Working Group on Materials (SWGMAT) that met under the aegis of the FBI laboratory, and she was on the earlier Committee on Forensic Hair Comparison. Both committees included international members. That is how she met her co-author, Jane Taupin.

Ms. Cwiklik is interested in the thought processes in forensic science, especially in trace evidence, and has published on the significance of trace evidence transfers and context-based examinations. She has a particular interest in heat damage to materials and has ongoing projects in that area. She has presented work involving fibers, hairs, debris, and thermal damage, as well as the significance of scientific evidence in the legal system, and has taught about forensic science to police officers, private investigators, lawyers, other forensic scientists, and the occasional university class.

Protocols, Procedures, and Philosophy

1

1.1 The Importance of Clothing Examination

The construction of clothing means that it can be a repository for a wide variety of useful information. Garments may retain various types of evidence that have been deposited onto them in a wide variety of ways, most importantly during the crime event. Clothing items are one of the most commonly encountered exhibits in crimes of violence. Generally, crimes are committed while the participants are clothed (although not always, notably some sexual offenses). Consequently, that clothing may reflect the nature, the location, or the participants in the crime. Clothing cannot literally speak to the examiner, so clothing is not "direct" evidence. However, garments may contain physical evidence relating to the crime and in an indirect way provide information relating to the circumstances. In this manner, we can consider that clothing "speaks" to the examiner.

Physical evidence must be detectable, whether by the human senses alone or aided by instrumentation or chemical means. It is tangible, so it can be analyzed and compared. The most critical aspects of any physical evidence analysis are (1) its physical detection and (2) its recognition as evidence relevant to the case (Inman and Rudin, 2001). Physical evidence from clothing includes most of the physical evidence that can be encountered in a crime. Blood, semen, fibers, hair, paint, glass, flammable liquids, and firearm discharges are examples. The material of the clothing item holds this physical evidence.

Biological evidence obtained from garments may be the most relevant evidence obtained in serious violent crimes, due to the innovation of DNA profiling. Semen from the offender may be found on the victim's underpants or other clothing and may be vitally important to identifying the offender if there is no medical examination of the complainant within a certain time frame. The presence of semen also indicates a sexual event. Blood may be transferred in a violent physical assault between the victim and the offender. Either of the participants' clothing may contain relevant evidence, such as the DNA profile of the blood, which may identify a person, or particular patterns of the blood (blood pattern analysis), which may assist in determination of the actions in the crime event. Clothing, if stored appropriately and not altered, may retain this evidence for many years and thus is extremely valuable.

Crimes such as robbery and burglary, which traditionally allowed for little biological evidence, may now provide evidence that links the offender to the crime. Due to the advances in DNA profiling, clothing found at a crime scene may be analyzed for "wearer" DNA. Areas of a garment in constant contact with the skin, such as the neckband and cuffs of a sweater or the nose and mouth areas of a mask, may yield DNA profiles that identify the wearer. This exciting DNA innovation is applicable to other crimes and illustrates another example of the importance of clothing examination. The following case study (*Queen v. Pike*, 2006), in which one of the authors was a scientific witness, shows how clothing may be used to identify an offender:

> Father and son Chinese restaurateurs in Liverpool, England, were severely assaulted in their home by three masked intruders. Video security cameras on the home captured footage of the

stolen getaway car, and the car was found abandoned some miles away. On the back seat of the car, authorities found a balaclava similar to that seen in the video footage. The balaclava was analyzed for DNA and yielded a profile that matched a young convicted offender on the national DNA database. This person was wanted in two counties for other offenses and was eventually located and charged. At the beginning of the trial, the defense requested the DNA evidence not be admitted as they believed the case reporting scientist was not in the country and thus could not be cross-examined on the DNA evidence. When informed that the biologist who did the analysis had just arrived in the country from Australia, the accused pleaded guilty.

Although not as specific to a particular individual as DNA, other physical evidence may also provide relevant information about the crime event and can provide strong evidence of contact or other activity. Hairs and fibers, fingerprints, footwear impressions, and even tire mark impressions may provide evidence of contact. Paint, glass, pollen, and soil may provide evidence of location. Gunshot residue may assist in shooting reconstruction. The presence and stage of growth of insects or larvae from the clothing of a deceased person may assist in the determination of time of death.

The examination of damage to clothing may provide information as to the possible implement causing the damage, the manner in which it was caused, and whether it was recent. Damage analysis may corroborate or disprove a particular crime scenario (Taupin, 2000). In addition to physical crimes between persons, any damage to the clothing may also assist in firearm-related events or fire scenes.

However, the most difficult challenge when presented with clothing is the recognition of the relevant physical evidence. We hope to describe in the following pages some tools and techniques for recognizing the physical evidence.

In addition to containing physical evidence, the appearance of the clothing may assist in the investigation of the crime. The form, color, style, or damage to the garment may support the observations of other professionals such as pathologists. The information may also support (or refute) the observations from the victim, the accused, or witnesses.

Data from additional observations, such as patterns of debris or blood deposits on clothing, may assist in determining if someone was wearing the clothing when it was in contact with other items, or if transfer was direct or indirect.

The information obtained from clothing may potentially have great evidential value, so it is mandatory to treat clothing items with due care and attention, from their first receipt to their role as court exhibits (and longer, if one considers appeals and retrials).

1.2 Clothing a "Crime Scene"

The examination of clothing may not only provide associative evidence with persons or locations, but also tell the examiner a "story" as to what happened during the crime event. We can even consider a garment or a number of garments from a crime as a "crime scene," albeit one that is transportable. Reconstruction of events from clothing examination may be considered similar to crime scene reconstruction. A crime scene may be cleaned once it is thought that all evidence has been obtained, and one cannot expect evidence to remain intact at that scene if it has been overlooked and examiners return days or even hours later. Clothing, however, can retain important evidence for years, even centuries. Many cold case homicides from decades ago have been solved by examining the stored clothing (or stored extracts from the clothing) and performing newer techniques, such as DNA profiling, that were not dreamed about in the days of the crime event. Just as vital to the criminal justice system are cases in which a convicted person is found to be innocent through DNA profiling not previously available. The case of *People v. Morin* in Canada precipitated a judicial inquiry into why Guy Paul Morin was convicted (Kaufman, 1998):

> A young girl's decomposed body was found outdoors in Ontario, Canada, in 1984, 3 months after she went missing. She was partially clothed, with her panties near her right foot, suggesting sexual assault, and she had been stabbed numerous times. Due to the body's state of decomposition, it could not be medically ascertained whether she had been sexually violated. Although Guy Paul Morin was arrested for and convicted of the murder, it was 10 years before DNA profiling could be performed on the semen found on the panties. The DNA profile from the semen did not match the DNA profile of Morin, and thus he was acquitted. A subsequent judicial inquiry uncovered many failings in the process of his conviction, including flawed forensic examination. It should be noted, however, that it was semen from the murdered girl's panties that still held the vital clue 10 years later. The body of the victim did not contain any foreign DNA, possibly because it was so decomposed. The clothing was still intact after 3 months and contained semen that could eventually be DNA profiled.

It should be noted that this case also exemplifies the limitations in the interpretation of physical evidence that may be present on clothing and the necessity to follow procedures to prevent contamination. Flawed hair and fiber evidence from the clothing helped to originally convict Morin (see Chapter 2).

1.3 Multiple Hypotheses, Alternative Explanations

The search for relevant evidence is made simpler by the formulation of competing hypotheses. Recognizing evidence relevant to a crime requires looking with purpose, rather than a cursory scan.

The scientist works by disproving alternative hypotheses. When the experiments to disprove alternative hypotheses cannot be conducted, when not all the alternative hypotheses can be disproved, or when the alternatives cannot be completely disproved, the forensic scientist reaches a qualified conclusion. One of the reasons qualified conclusions are more common in forensic science than in other scientific disciplines is that forensic science answers a different type of question: How did a unique event occur? rather than, How did a phenomenon occur? Another reason is that a decision must be made about a particular case whether or not all the test results are ideal, and the scientist must provide the best information possible at the time.

An investigator may develop a prime suspect during the course of an investigation, and thus will have a provisional hypothesis when processing a crime scene, for example. The investigator must be willing to update that hypothesis based on new information. However, the scientist should approach the clothing examination with multiple hypotheses rather than a single hypothesis, and before a final evaluation, he or she should consider multiple competing hypotheses. (This is discussed further in Chapter 9.)

A famous Australian case in which alternative hypotheses were not disproved was what became known as the "Dingo" case (Morling, 1987). The conviction of Lindy Chamberlain for the murder of her daughter Azaria prompted a Royal Commission, which discovered that flawed forensic evidence was presented at the trial. Lindy Chamberlain was freed after serving 4 years in prison:

> Baby Azaria disappeared from a campsite near Uluru (Ayres Rock) in central Australia; her body was never found. The prosecution alleged that damage to Azaria's jumpsuit, which was

found near the campsite after she had gone missing, was inflicted by either scissors or a knife. This supported their proposition that the mother, Lindy Chamberlain, had murdered her. In contrast, Lindy stated she "saw a dingo take my baby" and that the damage was caused by the dingo. During the trial, the prosecution alleged that dingoes could not produce damage resembling that on the jumpsuit. However, no one tested the hypothesis that dingoes could in fact produce similar damage. It was not until the Royal Commission of Inquiry into the convictions of the Chamberlains (Lindy's husband Michael Chamberlain was also charged and convicted as an accessory) that Lindy's hypothesis was tested. The principal scientist assisting the Royal Commission performed numerous experiments with dingoes on jumpsuits (Tony Raymond, personal communication). In his Royal Commission report, Justice Morling (1987) said, "it cannot be concluded beyond reasonable doubt that the damage to the clothes was caused by scissors or a knife or that it was caused by the teeth on a canid." The convictions were subsequently overturned.

1.4 The Origin of Evidence

A set of fundamental concepts applies to forensic analysis and consequently the examination of a clothing exhibit for evidence. The concepts of divisible matter and transfer relate to the generation of evidence (Inman and Rudin, 2002). Other concepts pertain to the recognition, analysis, and interpretation of evidence (Inman and Rudin, 2002). These concepts are identification, classification or individualization, association, and reconstruction.

Transfer of matter is often explained in terms of the Locard exchange principle, familiar especially to trace evidence analysts. This principle is often simply enunciated as "every contact leaves a trace." When examining the origin of evidence on clothing, one may consider this transfer on both a microscopic and macroscopic scale. Clothing from the presumed offender in a stabbing homicide may have the victim's blood spattered on it (macroscopic, showing proximity to offense) as well as fibers from the deceased's clothing (microscopic, showing possible contact). Different questions are answered with the two different types of evidence.

Divisible matter is that which divides into smaller component parts when sufficient force is applied. This is especially applicable to physical match evidence. However, divisible matter is generally not relevant to pattern transfer evidence, such as prints and impressions, because the pattern is the pertinent evidence, such as a greasy fingerprint on clothing.

Identification of the evidence defines its physicochemical nature (Saferstein, 1998). Identification is often the first process in the examination of an alteration to a garment. It answers the question, "What is it?" Classification of the evidence is the next step toward individualization. It answers the questions, "Which one is it?" or "Whose is it?" Class characteristics are general characteristics that separate a group of objects from a universe of diverse objects. Class characteristics can serve to screen a large number of items by eliminating from consideration the evidential items that do not share common characteristics with the group.

It is sometimes not possible to individualize a piece of evidence or infer a common source of origin, simply because the processes used are not capable. For example, a physical comparison of two hairs may classify the two hairs as being in a common group but will be unable to determine that they have a common origin. A DNA analysis of the roots of the hairs, however, may be able to "individualize" them to determine if they could have come from the same individual with a statistical degree of probability. Individualization means that the probability of the sample originating from another source is so low that the examiner can reach the opinion or inference that it is from the same source.

Individualization is the aim of all forensic identification science. The problem is that the various types of forensic identification science do not use the same scientific paradigm and do not report their conclusions in the same format (Broeders, 2006). DNA evidence is probabilistic and quantitative, whereas fingerprint evidence is a categorical match or non-match. Forensic identification procedures may lead to categorical elimination, but unless the number of potential sources is limited and known, no forensic identification procedure can lead to categorical identification (Broeders, 2006). This is a complex subject debated in the literature (Broeders, 2006; Stoney, 1991; Nichols, 2007).

Association can be described as a link — whether direct or indirect — between the source of the evidence and a target (i.e., the item of clothing) via materials on both items that may have a common source. The process of association involves competing hypotheses for and against the idea of a common source, and for and against the possibility of direct rather than indirect transfer. The significance of an association is especially difficult for forensic scientists to determine. The debate about the use of statistics in DNA profiling is a notable example.

Finally, a reconstruction is the ordering of the association of the evidence in space and time, or an understanding of the sequence of past events. A reconstruction of a crime scene is a large-scale version of this idea. We propose that examining clothing may be a smaller version of a reconstruction of a crime event. This reconstruction attempts to answer the "where, how, and when" questions asked in a criminal investigation.

1.5 Searching for Evidence and the Screening Effect

Searching for evidence involves a different thought process than interpreting evidence. Searching is a deductive process (Inman and Rudin, 2001), as a theory is formulated based on previous experiences and information. The examiner uses this working hypothesis to guide the search. As Inman and Rudin noted, the most difficult challenge in the investigative approach is the recognition of relevant physical evidence. If it is not recognized, it will not be examined and thus not interpreted. Successful searching requires presumptions, expectations, and preliminary hypotheses.

Police personnel or evidence technicians often first "screen" clothing exhibits for evidence before sending them to the forensic scientist for examination. During this screening process, they decide whether the garment may contain any useful information. Consequently, this is the stage where evidence is either recognized or not recognized, which ultimately reflects on the physical evidentiary value of the case. The term "screen" simplifies the searching process to that akin to a cursory glance. In reality, this is one of the most difficult aspects of forensic investigation, based on observation, hypotheses, and experimentation. Demoting the task of searching for evidence necessarily impacts on the subsequent meaning of that evidence.

Searching for evidence involves more than simply looking for a target evidence type. Examining a rape case, one would of course search for semen on the victim's clothing; semen is the target evidence that would denote sexual activity. But one would also consider other evidence types according to observation and hypotheses. The DNA profile of semen on the victim's clothing may confirm the presence of the suspect. However, if the court case is fought on consent, then the presence of semen, while establishing sexual contact, is irrelevant in addressing lack of consent. Damage to the clothing, however, may help to disprove or support the hypothesis of rape and the question of consent.

An often-quoted police phrase, "failure to search, failure to find," can be applied to clothing exhibits. One of the critical aspects of clothing examination is the observation or searching process. If the search for evidence is reduced to a "screen," then one of the most important phases of clothing examination is also reduced in importance.

1.6 Checklists, Guidelines, and Protocols

The terminology described below is used throughout this book for the examination of clothing. These should not be followed without question, but with consideration and allowance for a particular case. Clothing examination is necessarily nonprescriptive, and common methods or protocols may need to be adapted.

Method: The analytical basis for providing specific types of information; e.g., phase contrast microscopy, electrophoresis, infrared spectroscopy, STR-based DNA profiling
Procedure: Process for implementing one or several methods for a particular analytical problem
Protocol: Step-by-step details of a procedure
Guidelines: Reference points to follow in implementing a method or procedure
Checklist: A reminder list

Procedures and the attendant protocols should be selected to meet the demands of the case questions and the analytical questions that arise from them. We provide examples of the use of these terms in the following chapters, as they are all part of the process of clothing examination.

The FBI has sponsored a number of scientific working groups (SWGs) to produce documents outlining guidelines for qualifications, training, validation, testing, and report writing (Adams and Lothridge, 2000). Participants in these groups come mainly from U.S. government laboratories, although scientists from private and overseas laboratories have also been included.* The European Network of Forensic Science Institutes (ENFSI) also produces guidelines for forensic laboratories in Europe (Fereday and Kopp, 2003). These guidelines are not specific methods or protocols; rather, they direct that validated protocols in each laboratory be implemented.

The approach to casework should be influenced by a spirit of inquiry stimulated by the individual case circumstances, as well as by the narrower questions that each scientist can directly address within their respective disciplines. Consequently, the examination of clothing will incorporate a number of different methods according to what evidence is found on the garment, but the initial procedure of examination should be essentially the same. That is, general philosophical issues such as the history of the garment, why the garment was submitted, contamination issues, and health and safety will not alter. The methods used in the examination may alter according to the type of physical evidence present.

* The two authors of this book were on the SWGMAT committee from 1997 to 2003.

1.7 Nonprescriptive Holistic Approach

The holistic approach we describe in this text considers all elements of evidence recovered from the clothing examination and the how and the why of the relationships among those pieces of evidence. As well as a consideration of all the evidence, there should be a concordance of that evidence. For example, bloodstain pattern evidence should not contradict the ballistic analysis. Integration of the findings thus places the evidence in context.

An Australian rape case illustrates the consequences of a prescriptive, non-holistic approach to a forensic examination and the failure to consider all the items submitted to the laboratory in the context of the case (Queensland Court of Criminal Appeal, 2001):

> A 13-year-old girl was raped in 1999 in Australia. She initially denied knowing the rapist but then nominated Frank Button. Vaginal swabs from the victim yielded spermatozoa, but a DNA profile was not successfully obtained. Sheets and pillowcases from the girl's bedding were also sent to the same laboratory but not tested. Button was convicted and sentenced to 7 years imprisonment. After 10 months, his appeal was heard on the grounds that there was no scientific evidence presented at his trial. The laboratory then tested the bedding on insistence from Button's lawyers. The DNA profile from semen found on the bedding did not match the DNA profile of Frank Button. The laboratory then retested the vaginal swabs and obtained a DNA profile that again did not match Frank Button. Instead, the vaginal swab profile matched that of a convicted rapist. Frank Button was acquitted.

As we have mentioned, it may be useful to think of clothing items as crime scenes. A crime scene reconstructionist asks questions such as what is it, how does it fit, and what does it mean in context (Chisum and Turvey, 2006). So indeed should the clothing examiner. Each form of evidence must be in agreement with the other forms that are present. The choice of personnel to perform this reconstruction is problematic. When there is no scientist to integrate the findings of the evidence from clothing, all too often the police or the judiciary performs this task. How to address this problem is beyond the scope of this book, but we hope to prompt discussion and thought about this subject through the consideration of the multiple factors inherent in clothing examination.

1.8 References

Adams, D.E. and Lothridge, K.L., Scientific Working Groups, *Forensic Sci. Commun.*, 2(3), 2000.

Broeders, A.P.A., Of earprints, fingerprints, scent dogs, cot deaths and cognitive contamination—a brief look at the present state of play in the forensic arena, *Forensic Sci. Int.*, 159, 148–157, 2006.

Chisum, W.J. and Turvey, B.E., *Crime Reconstruction*, Elsevier Academic Press, Burlington, MA, 2006.

Fereday, M.J. and Kopp, I., European Network of Forensic Science Institutes (ENFSI) and its quality and competence assurance efforts, *Sci. Just.*, 43(2), 99–103, 2003.

Inman, K. and Rudin, N., *Principles and Practice of Criminalistics: The Profession of Forensic Science*, CRC Press, Boca Raton, FL, 2001.

Inman, K. and Rudin, N., The origin of evidence, *Forensic Sci. Int.*, 126, 11–16, 2002.

Kaufman, F., *The Commission on Proceedings Involving Guy Paul Morin*, Queen's Printer for Ontario, 1998. Available at http://www.attorneygeneral.jus.gov.on.ca; accessed April 2008.

Morling, T.R., *Royal Commission of Inquiry into Chamberlain Convictions*, Government Printer of the Northern Territory, Darwin, Australia, 1987.

Nichols, R.G., Defending the scientific foundations of the firearms and tool mark identification discipline: Responding to recent challenges, *J. Forensic Sci.*, 52(3), 586–594, 2007.

Queen v. Paul Pike, Liverpool Crown Court, Liverpool, England, January 2006 (details from author J.M. Taupin).

Queensland Court of Criminal Appeal, *Queen v. Frank Allan Button*, QCA 133, 2001.

Saferstein, R., *Criminalistics: An Introduction to Forensic Science*, 6th ed., Prentice-Hall, Englewood Cliffs, NJ, 1998.

Stoney, D.A., What made us think we could ever individualize using statistics? *J. Forensic Sci. Soc.*, 31(2), 197–199, 1991.

Taupin, J.M., Clothing damage analysis and the phenomenon of the false sexual assault, *J. Forensic Sci.*, 45(3), 568–572, 2000.

Preliminary Inquiries

2

2.1 Focus of the Examination

The focus of an examination depends partly on the stage of the case (Kind, 1994) and what is already known about the incident. Is it certain that a crime was committed? A decomposed body found in the woods or the desert may be someone who died of natural causes rather than by an act of violence. Can an examination of the clothing shed any light on the activities and circumstances surrounding the death? What has already been determined by autopsy? Are the police searching for a suspect? There may be clues on the clothing of the victim. If someone is already charged with a crime, does the evidence support the involvement of the defendant? At the trial stage, we need to know whether the defendant was involved in the crime itself, or if the links established to date with the victim, the scene, or even the incident may not be links with the actual crime.

The forensic scientist performing any type of examination needs to form an analytical plan. Regarding clothing examination, the first step in forming the plan is deciding what items to examine and in what order. A list of "what we know" and "what we don't know" is a useful starting point, followed by an assessment of what information would be most useful immediately (Cwiklik and Gleim, 2009). Some examinations may be urgent to assist police investigators in finding or deciding between suspects or scenes, whereas other examinations may not be needed until later in the case. In a hit-and-run vehicle–pedestrian collision, the most urgently needed examination is usually a search for paint chips or smears, which can provide clues to the vehicle that struck the person. A reconstruction of how the impact occurred, often crucial at the trial stage, need not usually be performed at the outset. However, if anything unusual is observed in the initial search for paint, the examiner may decide to follow up immediately. An example of a hit-and-run masking a murder is described in Case Example 4.6.

2.2 Information Concerning the Crime

The formulation of alternative hypotheses should be the foundation on which the forensic scientist searches for, and analyzes, evidence. Preliminary hypotheses can usually be formed before any evidence is opened, guiding the formation of the analytical plan. The scientist must have some idea of the circumstances of the crime in order to formulate the hypotheses. Even deciding what items to examine requires the scientist to know something about the case circumstances. This is not a trivial matter if several hundred items may have been collected from several victims and multiple crime scenes in a complex case, and it is important even when the scope of the case is more limited. The forensic scientist must have some information in order to direct the search for evidence on the clothing. Evidence must be interpreted in the context of the history of a sample prior to its collection and preservation.

One might assume that a clothing examiner would be biased by the information received about a case. However, the control for bias should be the scientific method, not restrictions on the exercise of science. Hypothesis formation and testing are at the heart of the scientific method, and formulating multiple alternative hypotheses at the outset is a powerful tool in controlling for the impact of potentially biasing information (Platt, 1964). If the information provided to the scientist is unduly restricted, then the impact of the scientific method on the case is diminished and the impact of witnesses, police, and attorney bias is increased (Cwiklik, 2006; Cwiklik and Gleim, 2009).

The more the scientist knows about aspects of the case affecting the physical evidence, the more directed their questions can be when examining an item of clothing. As long as alternative explanations or hypotheses for the evidence are posited at the outset, then examiner bias should be minimized. The information supplied to the scientist should also be relevant. Autopsy reports and victim statements are relevant; the results of identification parades are not.

Even if the scientist is presented with extraneous information, the scientific process need not be compromised as long as the scientific method is followed. (See Chapter 9 for further discussion of bias and controls for bias.)

The following information should be obtained by the clothing examiner:

- People involved in the incident under investigation
- Location(s) of the incident and any related sites or vehicles
- Sequence of events before and after the incident
- Time relationships
- Clothing worn by participants, which may include the victim(s), suspect(s), or witness(es), or even miscellaneous clothing from the scene
- Injuries to participants, including autopsy reports and sexual offense medical reports and relevant photographs
- Scene reports and photographs

2.3 Levels of Information

A clothing examination generally has three stages:

1. Observation of stains, deposits, and damage, including patterns and the general condition of the clothing. The data of the observations includes written and pictorial descriptions and measurements, where relevant.
2. Microscopic examination of stains, deposits, and damage, evaluation and comparison of patterns, and preliminary chemical testing.
3. Sampling of selected stains and deposits for further testing.

2.3.1 Description vs. Identification

A rigorous description entails observation and documentation of physical properties that can be used later to group materials or to decide whether testing is warranted. For example, even someone who is not a firearms examiner can report "apparent gunpowder particles," based on microscopic observation, and provide a record of it in the laboratory notes. If needed, another scientist can follow up with chemical testing to identify the components. Even someone who does not perform biochemical analysis can perform microscopy and chemical spot tests to describe possible semen stains; another scientist can test the stains later to identify semen.

2.3.2 Data, Results, Conclusions, Interpretations

Data includes observations and measurements, and conversely, observation implies data, whether it is qualitative or quantitative; a result implies a test. Results are objective conclusions regarding tests and observations. Conclusions are what the results mean. Interpretation entails relating results and conclusions to the case questions (Cwiklik and Gleim, 2009).

As an example, a reddish-brown stain is observed on the jacket of a suspect in a stabbing. This general observation is followed by microscopic observation of a reddish-brown glaze, typical of blood. The stain is tested for the presence of blood using chemical reagents that yield a certain color when applied to a sample of the stain. The presence of that color within a certain time limit is the test result. The conclusion is that it is positive presumptive for blood. "Presumptive" means that the test is indicative but not specific (i.e., a screening test). If the presence of blood is later confirmed and attributed to the victim, and the blood was deposited as spatter (a conclusion reached from observations of the shape and distribution of material in the blood droplets), one can further conclude that the suspect's clothing was near the blood of the victim while a force was applied to that blood.

2.3.3 Stability of Information

Data and analytical results should not change unless an error was made. However, conclusions may be affected by new scientific information, and interpretations are often affected by new case information. For many years, it was thought that analysis of trace elements in bullets via neutron activation analysis permitted the attribution of bullets with the same trace element profile to a specific batch. However, more recent research has shown this to be untrue (Mejia and Sample, 2002; Randich et al., 2004). Even though the data of a comparison performed earlier would be unchanged, a scientist could no longer reach the same conclusion from corresponding elemental profiles. New case information may be presented as a hypothetical question in court: the expert witness is asked to suppose that something is true and then apply the scientific information to that situation.

2.4 History of Exhibit

The clothing examiner should elicit and document the recent history of the clothing item; that is, who was wearing it, to whom it belonged (if not the wearer), whether it was purchased second-hand or used, where the clothing was located when it was collected as evidence, and, when possible, where it was immediately prior to, during the commission of, and after the crime. This places the clothing in a context so that material transferred onto the clothing can be evaluated for the possibility of contact with either the participants or the crime scene itself, and so that material transfer attributable to the incident under investigation is not confused with the normal background debris on the item.

The clothing examiner should be aware of any previous handling or examination of the clothing. Items may be inspected in the police evidence facility by attorneys or investigators. Some items are prescreened by police personnel before the scientist receives them. Every time the evidence is examined, opportunities for contamination occur, particularly when the items are being handled by personnel who are less aware of small particle evidence and trace DNA than a forensic scientist. Prescreening by nonscientist police personnel may involve circling stains that were visualized with an alternate light source, collecting obvious hairs, and spraying the clothing with reagents to visualize latent fingerprints. These actions may compromise the potential information value of the clothing, making any subsequent observations somewhat limited.

The clothing examiner must be familiar with the nature of various types of evidence, and how it may change and degrade with time and exposure to the elements. Clothing made from cotton or other plant material may be eaten by insects and animals or degrade with environmental exposure, whereas synthetic fibers exposed to the same environment may last much longer. Materials that degrade rapidly in hot and humid environments may last for decades, even centuries, in arid climates or in environments such as peat bogs. Hairs from Egyptian and Incan mummies can still be examined today, but hairs on a car windshield exposed to hot and tropical weather may be infested with microbial activity within days.

The increased sensitivity of DNA profiling means that the context of any DNA profile obtained needs to be carefully scrutinized. Underpants have been demonstrated to harbor sufficient sperm to produce a full DNA profile even after washing (Crowe et al., 2000).

2.5 Reference and Control Samples

The comparison of evidence samples to reference samples representing individual sources is a distinguishing characteristic of forensic science. The evidence samples are those obtained from extraneous material on the clothing items, or even the clothing itself. Any relationship of these samples to the crime event must be determined by testing. Sometimes the relevance of the samples to the crime event is not obvious and requires contextual study. This is especially true of trace evidence, but may be true of any type of evidence in the right circumstances. That is why samples of background DNA and background debris on a given item should be collected, as well as reference samples of particulate, fibrous, and biological debris from the environments of the usual wearer.

What are the appropriate reference samples? The reference samples — the known samples — must be from items or persons that are an inherent part of the incident under investigation (Stoney, 1987). Biological references, in particular DNA profiles, are not difficult to select, because they are collected directly from a specific individual rather than from the scene or the person's environment. This is not true of other types of evidence used to characterize a person, such as fibers from the person's clothing and any associated debris. Reference samples that represent the artifacts of a person rather than the person himself/herself must be the component materials of the artifact or must be considered as a set rather than as individual fibers or particles. The reference samples may not be unique or even distinctive, but they must truly represent what they purport to represent and not another object.

Samples from reference or known clothing items or individuals in a case should represent the scope of variation present in the item. For example, in evaluating whether foreign fibers on the clothing of a deceased person came from a carpet in the trunk of a vehicle, it is important to have a reference sample of the carpet that truly represents the carpet, including worn and stained areas of the carpet and even the binding or adhesive.

Control samples are a type of reference sample for test results — samples with the material of interest present and other samples with the material absent. The test reagent or process applied to these samples should yield positive and negative results respectively. Controls are used to ensure that a particular scientific test is functioning correctly and are thus part of quality control.

2.6 Preservation, Handling, and Storage

It is not possible to separate collection and short-term preservation of a sample or item. Once a stain is collected, the evidence is altered from its original form. Thus, preservation issues should be considered in determining the collection method. One of the main considerations is whether the evidence can degrade, decompose, evaporate, or change. Biological evidence, whether body fluids or plant material, is subject to degradation. Clothing items containing biological evidence should be dry (if wet, then air-dried in a fume hood) and stored in clean paper bags in a dry environment. Biological samples such as liquid blood should be placed in a −4 °C freezer for long-term storage.

Labile evidence, such as flammable liquids, requires that the clothing be packaged in airtight containers. If an item contains several types of evidence with conflicting storage requirements, examiners from the respective specialties should immediately confer. It may be necessary to photograph and cut out selected body fluid stains and place the sampled clothing in an airtight container to allow testing of flammable liquids. Clothing with deposits of acids or other chemicals that may continue to react with the substrate should be examined immediately and the stains excised to contain the damage. If examination is delayed, neither the chemical nor the clothing may remain in its original condition.

Clothing that is wet needs special consideration because it may contain nonbiological as well as biological evidence. Once mold or fungal material appears on the item, any evidence is severely compromised, if not destroyed. Clothing should be air-dried as soon as possible in fume hoods and with regard to any possible contamination from other items or indeed the fume hood itself. Air flow around the clothing item should be low enough to prevent removal of any loose fibers or particles adhering to the item. A clean piece of butcher paper should be placed below the item to collect any debris that may fall off. Clothing with wet stains should be folded in such a manner that prevents transfer of the stains and deposits to other parts of the garment. If necessary, the item should be wrapped in butcher paper so that the stains do not transfer and their integrity is maintained as much as possible.

Clothing items should be submitted to the laboratory in separate packages and under separate item numbers. Packages should not run the risk of breakage, spoilage, or contamination. Bottles and other glass or ceramic items should not be packaged near sacks containing clothing, because of the potential for breeching the sacks and damaging that clothing.

It should go without saying that the proper storage of exhibits is essential. Laboratory evidence storage is usually limited to cases that are awaiting analysis or are waiting to be returned to the submitting agency, usually a police or sheriff's department or other government entity. If police evidence storage is inadequate, laboratory management should assist by providing storage requirements for physical evidence, including special requirements for specific types, and lending their support to police funding requests for upgrading their facilities accordingly. In our experience, police departments usually want to do the right thing when storing evidence and are taking the steps to do so. The notable exceptions illustrate the consequences of failing to follow through (for example, see Bromwich, 2005).

2.7 Contamination Issues

Improper preservation, handling, or storage may result in the introduction of extraneous substances to the garment — that is, contamination. Contamination of clothing items has severe consequences to the integrity of the evidence, to the examinations, and thus to the case itself. Contaminants may be introduced at the scene, during transportation, at the autopsy or hospital, or in the laboratory. Cross-contamination of clothing items from the victim and the suspect would be devastating to the case. There have been several prominent cases in which cross-contamination of DNA (Thompson, 2003; Johnson, 2006) and of fibers (Kaufman, 1998) led to apparent links with a victim or suspect that were in fact in-house contamination and linked only through the laboratory. The government entities involved in these two cases instituted inquires, in one case after a false conviction, and the laboratories have since evaluated their internal procedures.

Reference samples and evidence samples from the clothing items should be kept separate in time and space. That is, evidence samples should be examined in a different area and/or at a different time from reference samples. Modern DNA laboratories now completely separate the processing of reference and evidential samples, sometimes including different personnel and different instruments. A similar mindset should also apply with other items or evidence from clothing. It is imperative that protocols and procedures for preventing contamination be followed faithfully, so that there is little or no risk of contamination, and when it does occur, it will be duly recognized by personnel and documented.

The evidence must also be protected from the examiner, who should wear a face mask while facing evidence so that tiny droplets of DNA-bearing saliva are not transferred. Hair should be controlled so that it does not fall onto an evidence item. Protective clothing, including gloves, prevents transmission of debris from the examiner's clothing and person to the evidence being examined.

2.8 Health and Safety

The examiner must also be protected from the evidence. There is inherent danger in examining clothing items with deposits of body fluids and occasionally other artifacts from a crime, such as weapons or poisons. If the garment is still wet, then there is a possibility of inhalation of spores or aerosols and thus working in a protective chamber or fume hood is mandatory. Pockets or other cavities in the clothing may conceal weapons.

Wearing protective clothing, using face masks, allowing the clothing to dry completely, and other measures that appear in laboratory guidelines should be strictly followed. Laboratory protocols regarding health and safety must be followed at all times. The clothing examiner should always approach a garment with due caution and an awareness of the potential danger.

2.9 References

Bromwich, M.R. *Third Report of the Independent Investigator for the Houston Police Department Crime Laboratory and Property Room*, June 30, 2005. Fifth report, June 13, 2007. Available at http://www.hpdlabinvestigation.org; accessed April 2008.

Crowe, G., Moss, D., and Elliott, D., The effect of laundering on the detection of acid phosphatase and spermatozoa on cotton T-shirts, *Can. Soc. Forensic Sci. J.*, 33(1), 1–5, 2000.

Cwiklik, C., *A Good Look at Blind Testing: Quality Assurance Systems and Bias*, American Academy of Forensic Sciences meeting, Seattle, WA, February 2006.

Cwiklik, C.L. and Gleim, K.G., Forensic casework from start to finish, in *Forensic Science Handbook, Vol. III*, Saferstein, R., Ed., Prentice-Hall, Englewood Cliffs, NJ, 2009.

Johnstone, G., *Coroner's Report into the Death of Jaidyn Leskie*, Melbourne, Victoria, Australia, October 2006.

Kaufman, F., *The Commission on Proceedings Involving Guy Paul Morin*, Queen's Printer for Ontario, 1998. Available at http://www.attorneygeneral.jus.gov.on.ca; accessed April 2008.

Kind, S.S., Crime investigation and the criminal trial: A three chapter paradigm of evidence, *J. Forensic Sci. Soc.*, 34(3); 155–164, 1994.

Mejia, R and Sample, I., Chemical matching of bullets fatally flawed, *New Scientist*, April 17, 2002.

Platt, J.R., Strong Inference, *Science*, 146, 347–353, 1964.

Randich, E., Tobin, W., and Duerfeldt, W., Cited in: Forensic analysis: Weighing bullet lead evidence, *Report by the Committee on Scientific Assessment of Bullet Lead Elemental Composition Comparison, Board on Chemical Studies and Technology*, Division of Earth and Life Studies, National Research Council of the National Academies, National Academies Press, Washington, DC, 2004.

Stoney, D.A., Fundamental principles in the evaluation of associative evidence, *Can. Soc. Forensic Sci.* (Special Edition: Abstracts of the 11th meeting of the International Association of Forensic Sciences), 20(3), 280–282, 1987.

Thompson, W.C. *Victoria State Coroner's Inquest into Death of Jaidyn Leskie*, December 3, 2003. Available at http://www.bioforensics.com; accessed April 2008.

Preliminary Assessment 3

This chapter describes procedures for the initial examination of clothing items once received into the laboratory. Detailed examination for specific types and sources of evidence contained on and within garments are described in Chapters 4–8.

Documentation of clothing items is described here, followed by processes for the preliminary searching, detection, and recovery of evidence on garments. Alterations to the original condition of a garment are discussed. Clothing construction, including fabric and yarn composition, follows. Finally, common terminology used in sewing and manufacturing of garments is presented in Sections 3.9 and 3.10.

3.1 Documentation

The standard for documentation is that another examiner should be able to replicate the results from the notes and that the same or different examiner should be able to understand the evidence and the testing, even many years after the original work was done and after methods have been updated and common practices have changed. Consequently, documentation is very important and crucial for the examiner, the reviewer, and ultimately the court system. The clothing examiner may also be working under laboratory guidelines, especially if the laboratory is accredited and has accredited methods and training programs. The laboratory accredited general protocol should be used as a guide, but it should also be remembered that each criminal case examined is different and may have its own specific challenges.

Continuity, or the chain of custody, of the clothing item should be explicit. Any gaps should be noted, and subsequent limitations in the conclusions due to lack of continuity should be explained. Once the garment is received into the laboratory, all movement and all personnel examining the item should be documented. This is so that any tests and results may be fully explained to other scientists and subsequently the court. This is also as much a "history" of the garment as that prior to its receipt into the laboratory, as described in Chapter 2. Removing samples from the garment for testing will affect the condition of that garment; similarly, so may handling fragile garments, wetting with water for semen screening, specific types of storage such as freezing, and so on. Continuity labels should be described (written or photocopied) in the notes, including the description of the item and dates and times received. The garment should be described exactly as received. If the garment is received inside out, folded while still damp, or with other garments packaged together, then these points should all be noted because they may have affected the original condition of any evidence.

The item of clothing should be described in terms of the type of garment, where it was worn on the body, color(s), composition of yarn(s) in the fabric, and type of fabric. Garments are most often worn as designed for a particular purpose; that is, a T-shirt is usually worn on the body covering the upper chest area. But in criminal cases, one may come across T-shirts that have been placed in the mouth as gags, used to staunch bleeding wounds, or fashioned to form a head covering in armed robberies. The circumstances of the case should thus always be borne in mind when describing the type of garment.

Additions to the body of the garment, such as pockets, collars, bows, trims, appliqués, etc., should be noted because they are distinguishing features. Any labels attached to the garment should also be described, which may include the size, make, and fabric composition. Many laboratories have proformas for items of clothing that include basic diagrams, such as T-shirts, long-sleeved windbreakers, and jeans. These can simplify the process as long as it is recognized that modifications will need to be made. Although it is not necessary to make every diagram to scale, it may be of value to incorporate scales in diagrams with pertinent garment measurements (i.e., length and width of sleeves, length of trouser legs), and it is important to include measurements of bullet holes and stab cuts and any other defects that can be correlated with information about injuries in autopsy and medical reports.

Photographs are an especially useful tool for an examiner's notes and are quick and easy to take now using current digital cameras. Overall shots of the front and back of the garment should be taken. Close-ups can be taken of various pertinent areas once recognized. Again, scales should be included in any photograph. Such photographs do not substitute for a good diagram and vice versa, as both emphasize different information.

Appendix 1 summarizes the types of features and documentation that should be noted in a preliminary examination of a clothing item.

3.2 Detection

Clothing items should be laid flat on a clean surface and systematically searched. The use of butcher paper underneath the clothing item and on top of the laboratory bench will enable the collection of any loose debris. A mannequin or model may be helpful for a three-dimensional perspective and the orientation of the clothing. A mannequin is especially useful in damage or blood pattern analysis cases and when a number of garments that have evidentiary value are worn by the one person.

The first examination is an unaided visual examination; that is, with the naked eye. A wide array of physical evidence can be observed in this manner. Staining, patterns, deposits, and damage may be readily discerned. Light-colored clothing in lightweight fabrics, such as a white cotton polyester business shirt, may require only this step.

Appropriate lighting, such as oblique lamps, magnification, or alternate light sources, may be necessary for dark-colored fabrics or heavier fabrics. Infrared and ultraviolet light sources have been used in published case studies. Ultraviolet light was used to better visualize bloodstaining on the defendant's brown corduroy pants in a "stomping" case (Ristenbatt and Shaler, 1995). Infrared imaging has been used recently to record latent evidence such as bloodstains and gunshot residue (Lin et al., 2007). In fact, 20 years ago infrared photography was used to elucidate bloodstain patterns in an Australian case (Raymond and Hall, 1986).

Light sources and photography should always be considered when the staining (such as blood) is difficult to discern because of the background color of the clothing (dark red or black being particularly difficult in blood cases). These techniques are nondestructive, simple, and cost-effective, presuming the laboratory has the appropriate equipment. Figure 3.1 shows two photographs of the left leg of a dark-colored (almost black) pair of jeans. In the bottom photo, the passive drips of blood are difficult to discern in visible light. However, the top photograph demonstrates the use of infrared light, allowing the drips of blood to

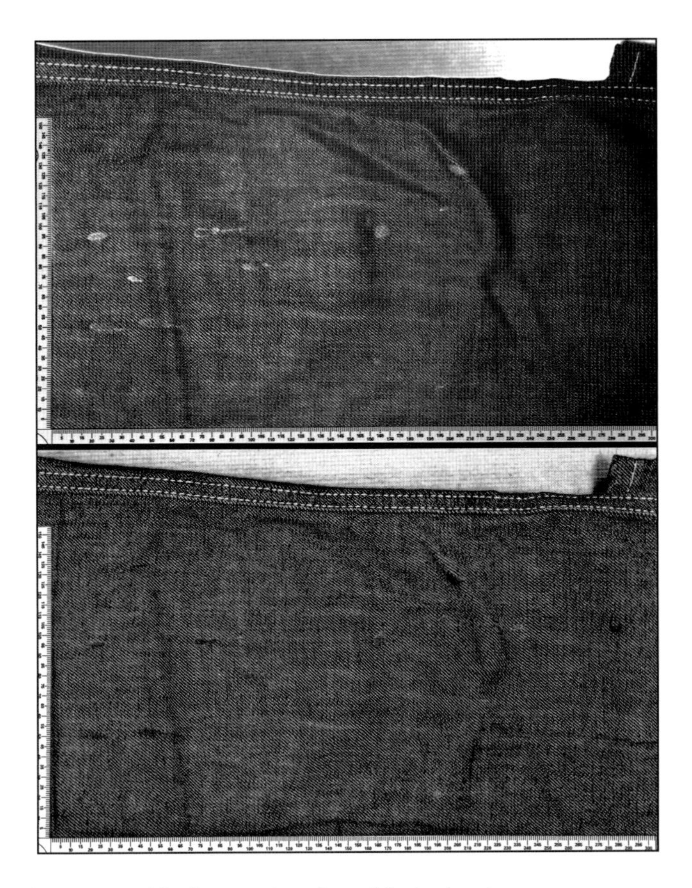

Figure 3.1 Leg of jeans: visible (bottom), infrared light (top).

be readily seen. The photograph on the left in Figure 3.2 shows a mannequin wearing a black sweater placed under visible light; for the photo on the right, infrared light was used to illuminate the room of the laboratory. Blood spatter was very difficult to elucidate in visible light, but infrared light allowed the morphology of the stains to be readily observed. This enhancement is useful not only in the actual detection of the bloodstains but also in photography for permanent records.

Biological stains other than blood may also be difficult to detect on clothing and the use of a versatile light source may assist in their detection. A light source composed of varying wavelengths may be used to locate potential semen, saliva, and bloodstains, especially when these are not visible to the naked eye. It may be used for the rapid detection of semen stains, rather than time-consuming (and possibly sample-consuming) chemical screening tests. It may also be used for the enhancement and photography of untreated bloodstains (Stoilovic, 1991). It is a relatively safe, simple, and nondestructive technique that may be used at crime scenes as well as on garments in the laboratory. Most variable light sources allow selection of wavelengths from ultraviolet (350 nm) to red (600 nm).

The color and type of the material of the garment on which a stain is deposited may have an effect, however, on the detection of that stain using a variable light source (Vandenberg and van Oorschot, 2006). The examiner should be aware of the background color of the garment, its potential fluorescence, and the appropriate excitation/emission conditions in order to enhance the contrast between the stain and the background (Vandenberg and van

Figure 3.2 Sweater: visible (left), infrared light (right).

Oorschot, 2006). The search of clothing with a variable light source is best performed in conjunction with other screening tests to better target an area of interest.

A low-magnification stereomicroscope (a low-power microscope that yields a three-dimensional image) follows unaided examination and is used to locate as well as investigate the nature of the physical evidence. A quality stereomicroscope capable of various forms of illumination, such as transmitted light, dark field, and axial and oblique reflected light illumination, is valuable, but most clothing examination can be performed with a good reflected light stereomicroscope with oblique illumination. Photography will ensure a permanent record of the image. A photographic attachment to the microscope is useful, but good photographs can also be obtained simply by placing the camera lens on one of the eyepieces. [See Appendix 2 for further information regarding stereomicroscopes. Other microscopic techniques and instrumentation may be found in specialized texts (e.g., Petraco and Kubic, 2003).]

3.3 Recovery

As a general rule, the simplest technique is the best for recovery of physical evidence from a garment. Hand picking with clean tweezers is suitable for trace evidence such as hairs and fibers and will be further discussed in Chapter 8. Lifting with tape lifts will recover a larger amount in a shorter time but will also gather extraneous material. Scraping, combing, and clipping are other techniques, but they may be more destructive to the garment. (Please see discussions in Chapters 4 and 8.) Stains are usually cut out from the material to determine the presence of the components, such as spermatozoa or blood, but enough stain should be left to allow a second examiner to determine how it got there (i.e., deposited as a film, crust, from the inside/outside).

When clear tape is used to collect evidence, the tape is searched using a stereomicroscope. The background chosen depends on the color of the material to be located. For example, dandruff is viewed over a dark background. The dandruff material is removed by cutting a small window around it with a scalpel and removing the tape along with the attached piece of dandruff into a sterile tube, suitable for DNA samples.

It is not necessary to sample or recover any or all physical evidence on a garment. Patterns, such as blood patterns or impressions, may be left on the clothing for interpretation and subsequent reexamination. Sampling for DNA analysis or other composition may be required, but this should be done with minimal disruption to any pattern or the garment in general.

3.4 Clothing Construction

Determination of the origin of items of clothing may be requested by the investigator. A garment may be found at a scene associated with a crime, and police may suspect it was worn by a victim or an offender. The type and size of the garment is often the easiest question to answer and can be furnished generally without prolonged examination. A brassiere is most likely to have been worn by a woman (unless in a gender change case), and men's trousers with a long inner leg seam are most likely to have been worn by a tall man. Any distinctive or unusual features on the garment may also be promptly given to the police investigator. During investigations of missing or kidnapped persons, the families or friends of the individual may quickly recognize a distinctive garment that has been located by the police. Sometimes a label with the name of the owner is sewn to the garment if the wearer lives in an institution, is a child, or is mentally infirm.

The advent of closed-circuit television (CCTV) footage in buildings and streets has often enabled investigators to determine the type of clothing worn by an individual on the day of a suspected crime or event. A woman in Victoria, Australia, alleged that she had been attacked in bushland by an unknown offender; her damaged top was submitted for examination (Taupin, 2000). There were numerous recent scissor cuts in the top, which contradicted her story of the offender manually tearing the garment. Furthermore, when the detectives were provided with a full description of the garment, they realized it was completely different to that observed on the CCTV footage of the woman in a shop on the day in question, shortly before the alleged incident. It was then postulated that the woman had cut another garment at home to mimic the assault. The woman subsequently pleaded guilty to false report, admitting she had wanted to obtain money from crimes compensation legislation.

Labels on the garment should be noted; information includes the fabric composition (for example, cotton), the size, the country of manufacture, and the brand. The manufacturer's label on the garment (if present) may help to identify the wearer's build and shape. Many labels have the size of the garment imprinted on them. However, caution should be used, because these sizes can vary from brand to brand and from country to country. Construction size tables vary from Europe to the United Kingdom, the United States, and Australia. A size 10 woman's dress in Australia is a size 6 in the United States and a size 38 in Europe.

The construction size tables are important for tailors during the manufacture of clothing. These tables specify the dimensions of a garment and provide measurements that are used for cutting out the component pieces. It is not unusual for 50 or more layers of material to be cut at the same time (Adolf, 1999). This mass cutting often means that one garment is sewn together from different batches of fabric. The result is that a single item of clothing may consist of fabric dyed in several batches, an important fact to remember when taking control samples. The stitching together of the fabric pieces by the tailor involves the formation of seams. Some seams may be stitched to be stronger than others and may become important in damage analysis. [For further terminology, see Section 3.10. A comprehensive reference book for terminology in the clothing manufacturing industry may be a useful text in the laboratory (Textile Institute, 2002).]

The following case study illustrates salient points when attempting to identify the owner or wearer of a garment submitted to the laboratory for forensic examination:

> Neighbors reported a disturbance outside their homes late one night in northern England during 2009. They heard shouting and a man's voice saying "please don't kill me" and then the sound of a car speeding off. A bloodstained, damaged T-shirt was located in the vicinity as well as a pair of bloodstained training shoes. The examining forensic biologist was able to quickly supply the size and description of the clothing items to the investigating police. Damage to the T-shirt was then analyzed, providing information as to the type of assault — extensive recent tearing in the garment but no cuts or bullet holes suggested the use of manual violence only. The distribution and morphology of the bloodstains to the items accorded with this theory. Subsequent DNA analysis of the bloodstains and of appropriate areas on the T-shirt and trainers for "wearer DNA" yielded three separate DNA profiles. A search on the National DNA Database provided three names; each individual was located and interviewed. It appeared that the event was a dispute over drug transactions. No charges were brought and the case was closed. The clothing examination was not only able to locate the parties involved but was also able to identify the type of assault. Most of the steps in the clothing examination were relatively timely (in the order of hours only), with the longest being the DNA analysis (2 days). Consequently, the police were quickly able to disband a potential murder investigation and allay the public's fears using the results from the clothing examination only and without traditional on-the-beat detective work. (case analyzed by J.M. Taupin)

Obtaining wearer DNA from a garment may not necessarily resolve the issue of who was wearing the garment at the time the crime was committed. Fresh footprints in the snow leading away from a home burglary led patrolling police to a town square. A man in the town square was acting suspiciously, and items from the burglary were found on his person. When questioned by police, the man said his brother did the burglary and then forced him to wear his shoes and take the stolen items from him in the town square; the man said he had never been to the house. The relevance of who wore the shoes was thus reduced because the man was apprehended wearing them. The size of the shoes was more relevant; it appeared the shoes were ill fitting. (further information from J.M. Taupin)

Sometimes incomplete pieces of material (decomposed, damaged, or burnt) may have a characteristic such as a zipper or buttons that enables the scientist to suggest that they came from a particular garment. A skeleton was found down a mine shaft in Victoria, Australia, and remnants of the accompanying clothing were submitted for examination. Synthetic stitching threads and numerous metal studs with adhering small sections of denim yarn suggested that they originated from a denim jacket. Denim, which is made of cotton, is susceptible to microbial degradation (see Chapter 6) and would have disintegrated far more readily than the synthetic stitches. (case study from J.M. Taupin)

The term "textile" is applied to fibers, filaments, or yarns, natural or man-made, and products made from them. Threads, cords, ropes, braids, lace, embroidery, nets, and fabrics made by weaving, knitting, felting, bonding, and tufting are all textiles.

The basic starting material for the manufacture of a textile is a fiber, which may be either a staple fiber or a filament.

3.5 Yarn and Fabric Composition

The clothing examiner should have some knowledge of the structure of different kinds of textile fabrics, such as weaves, knits, and non-wovens, and also an understanding of their mechanical and physical properties, such as elasticity. It is not essential that the clothing examiner become an expert in textile composition and construction, but a basic understanding is strongly advised. This knowledge would also assist the examiner in understanding deposition, transfer, and persistence and other fiber/yarn/fabric theories. The yarns used to make the material of the garment may also influence damage characteristics. Visits to textile manufacturers are useful in order to understand the processes involved in making textiles, such as weaving and knitting.

3.6 Yarns or Threads

Generally, clothing today is produced from yarns, such as woven and knitted fabrics. Laces and nets also fall into this category. Fabrics made directly from fibers without yarns include felt.

Thread is a generic term meaning textile yarn and is the elementary unit of textile construction. Thread forming, which starts with a spinning process, is a basic step for the manufacture of most textiles. The most common forms of yarns are single yarns and multiple-wound yarns, which are two or more single yarns wound together parallel.

Single yarns are produced either from staple fibers (fibers of limited and relatively short length) or from filaments (fibers of indefinite length). Examples of natural staple fibers are wool, cotton, and jute. Filaments are fibers of indefinite length. Examples of filaments are silk (a natural fiber), polyester, and nylon.

A single yarn consisting of staple fibers is described as spun yarn. Spun yarns are usually held together by twist. An essential feature of spun yarn is a greater or lesser degree of hairiness. A single yarn composed of one or more filaments is referred to as filament yarn. Filament yarns consisting of more than one filament are known as multifilament yarns. There are multifilaments with and without twist.

Folded yarns include all threads in which two or more single yarns are twisted together in one operation. Depending on how many yarns have been twisted together, there are two-folded yarns, three-folded yarns, etc. Folded yarns used in hand knitting are described as two-ply, three-ply, etc., and this description of yarn is used in this book.

Manufactured fibers can be produced as a staple or a multifilament yarn. The method of tow conversion (cut or stretch broken) determines the fiber end appearance of the constituent staple fibers. Constituent fiber ends have been defined as the fiber appearances produced through the manufacturing processes of tow-to-staple conversion.

The concept of "hairiness" in a yarn is important also for trace evidence transfer of fibers. Sheddability, fiber transfer, and fiber persistence are principally associated with the hairiness of a fabric. The hairiness is generally determined by both the construction of the fabric and the construction of the yarns used for its manufacture. Trace evidence transfer is discussed in Chapter 8.

More information regarding the formation and analysis of yarns can be found in SWGMAT Guidelines or in other texts (e.g., Robertson and Grieve, 1999).

3.7 Fabric

A fabric is a manufactured assembly of fibers and/or yarns that has substantial surface area in relation to its thickness and sufficient inherent cohesion to give the assembly mechanical strength.

3.7.1 Weave

A weave is a fabric manufactured on a loom and composed of regularly interlaced threads or yarns. These threads are called the warp threads and the weft threads. The *warp* is oriented lengthwise in the fabric, whereas the *weft* is introduced widthwise (Figure 3.3). The individual warp yarns are called the ends and the individual yarns in the weft are called picks. The pattern of the interlacing of the warp and weft yarns is known as the weave; the repeat unit of the weave contains the smallest number of ends and picks that adequately represent the interlacing pattern.

There are three basic types of weave: the plain weave, the twill weave, and the satin weave (see Figure 3.4). The plain weave is the simplest and most frequently used in clothing. It can be considered as "one over and one under." The plain weave, because it requires less fiber mass, is often used in ladies' blouses, business shirts, and some trousers.

A twill weave produces diagonal lines on the surface of the fabric and is considered a high fiber mass. Consequently, their durability and resistance is higher than plain weaves, and so twill weaves are often found in clothing used as workwear. The most common kind of twill weave is denim, used in jeans and jackets. Men's gabardine suit trousers are also twill weave fabrics.

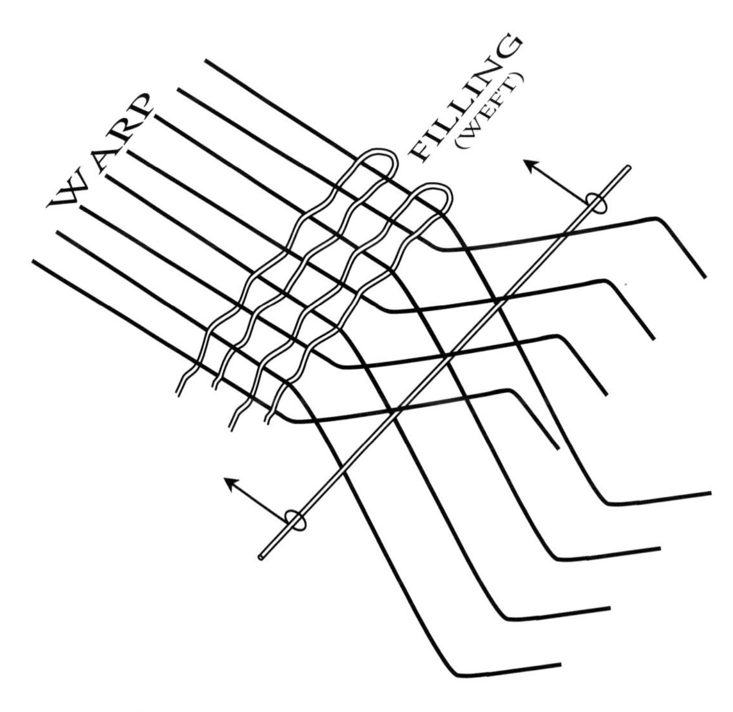

Figure 3.3 Formation of a weave.

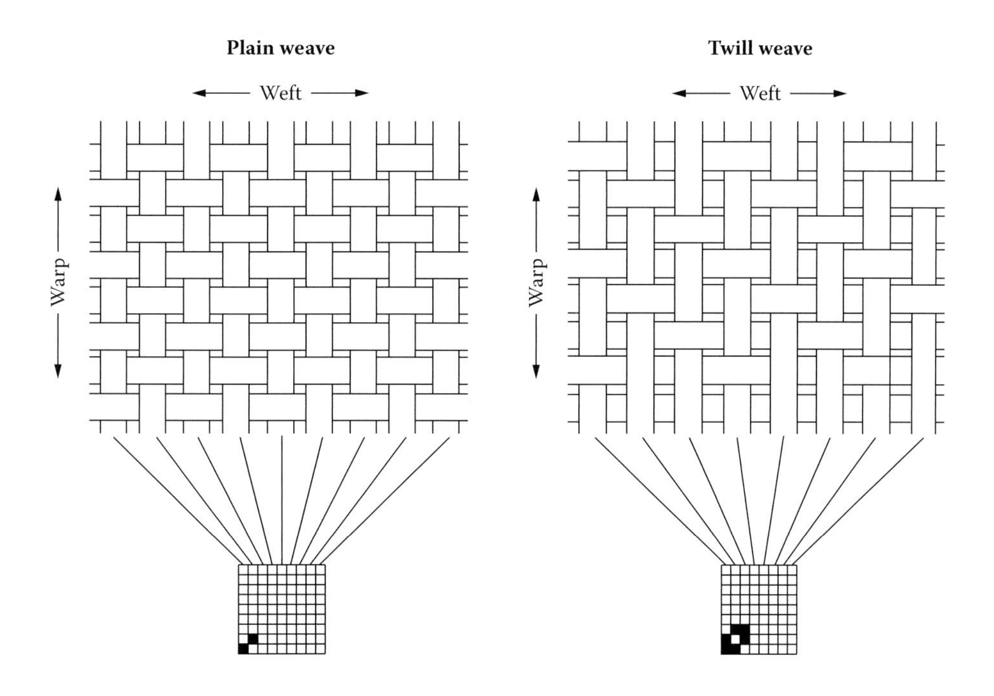

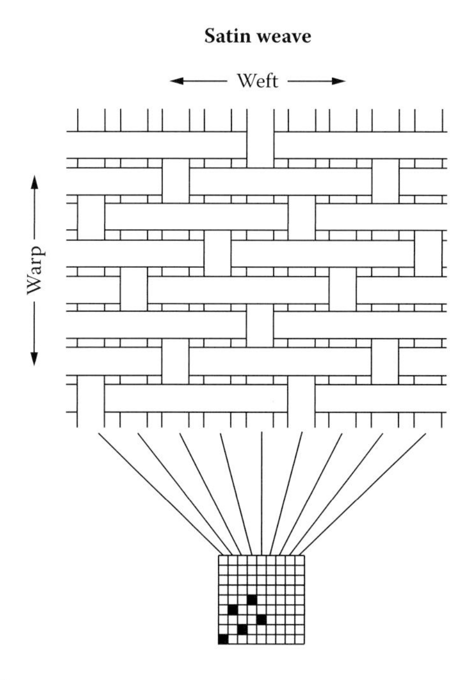

Figure 3.4 Basic types of weave.

Satin weaves have a smooth surface and are either weft-faced or warp-faced fabrics. These are called sateen/weft or satin/warp. They are often found in fine table linen as well as in clothing.

3.7.2 Knit

A knit is a fabric either knitted by hand by needles or on power-operated knitting machines. Knitting is a process that intermeshes loops of yarn (Figure 3.5). The diagram in Figure 3.5 shows the face of a plain jersey knit that is composed completely of face loops and one that is composed completely of back loops. Loops are formed by needles and then new loops are drawn through those previously formed — the continuing addition of new loops creates the fabric. Knitted loops are sometimes referred to as stitches when they are pulled through another loop. Consequently, yarns in knitted fabrics follow a three-dimensional configuration, contrasting with the yarn configuration in woven fabrics, which is essentially two dimensional. Loops running in the machine direction are known as wales (analogous to the warp direction in a woven fabric) and those in the cross-machine direction are known as courses (Figure 3.6).

Weft knitting, regarded as normal knitting, has loops that are made by one single thread at a time. The weft threads are fed more or less at right angles to the direction in which the fabric is produced. Hand knitting is a weft knitting procedure. Warp knitting forms loops with many parallel threads at the same time. T-shirts, sweaters, and undergarments are usually knitted.

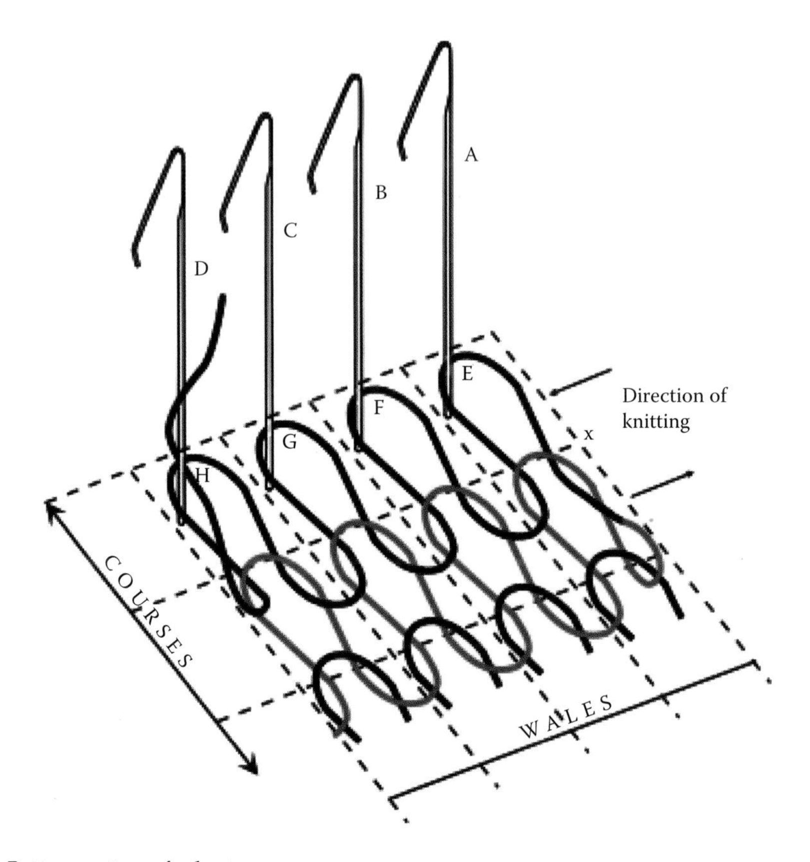

Figure 3.5 Formation of a knit.

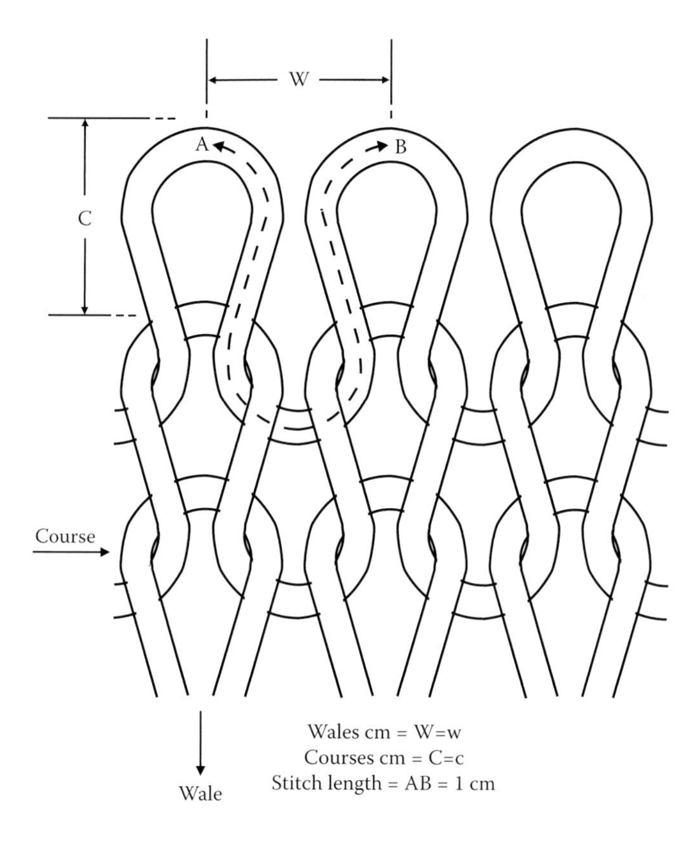

Figure 3.6 Courses and wales in a knit.

The basic types of weft knitting are plain knit, rib, and purl structures. Plain knit fabrics, also known as single jersey or single knit, have all loops drawn to one side or face of the fabric (Figure 3.7). A wide variety of knit fabrics are made with single jersey construction, ranging from sheer lightweight hosiery to T-shirts and bulky sweaters.

Rib knit fabrics have loops in alternate groups of wales meshed on one side of the fabric, with the loops in the remaining wales meshed on the opposite side. Each wale contains either all plain stitches (also called face stitches) or all purl stitches. Due to the greater extensibility in the coursewise direction, rib fabrics are popular as garment body and sleeve borders (e.g., waist, cuff, and wrist) as well as socks.

Purl knit fabrics have loops in alternate groups of courses meshed on one side of the fabric, with the loops in the remaining courses meshed on the opposite side — each wale contains both plain stitches and purl stitches. The appearance of the face and back are identical, and the fabric is highly extensible in all directions. Their high extensibility has favored their use in children's wear.

3.7.3 Felts, Leather, and Other "Non-Wovens"

Felts are produced by the compression of single fibers; leather derives from the skin of an animal. We group them together because they produce similar characteristics when damaged. Leather jackets are popular today, and leather and felt are common garment trims. Other non-woven fabrics may be composed of randomly oriented synthetic fibers held together by

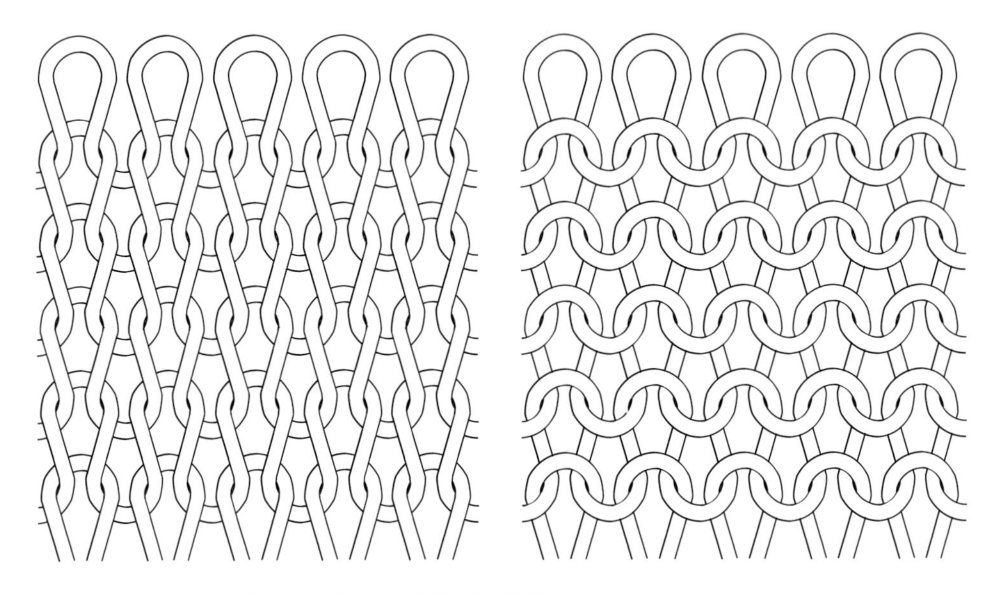

Figure 3.7 Structure of knits, front and back of plain jersey.

small drops of polymer, either from melting of the parent fibers or by addition to them. These are found in clothing primarily as facing or other layers beneath the surface.

3.8 Definitions

Fabric: Manufactured assembly of fibers and/or yarns
Fiber: A unit of matter characterized by flexibility, fineness, and a high ratio of length to thickness
Yarn: Elementary unit of fabric construction; a long and thin textile product generally less than 4 mm in diameter

3.9 Sewing Terminology

For weaves:

Bias: Any diagonal intersecting the lengthwise and crosswise threads
Grain: The direction in which the threads composing the fabric run
Plain weave: Simplest and most common of weaves; threads in both warp and weft direction run alternately over one and under one of the yarns they cross (e.g., gingham, linen, calico)
Satin or sateen weave: Surface composed of floats or warp thread that pass over many weft threads (four or more) before being caught under one; the weft threads are obscured and the fabric has a characteristic luxurious shine
Selvedge: The narrow, flat woven border resulting at both lengthwise sides when the crosswise threads reverse direction
Twill weave: Fabric in which weft threads pass alternately over one warp thread and under two or more, thus producing diagonal ridges (e.g., denim and gabardine)
Weave: Fabric formed by interlacing two sets of yarns at right angles

For knits:

Course: A row of loops along the length of a fabric
Crochet: Warp knitting
Double knit: Method utilizing a double knit mechanism in an ordinary knit, giving the back and front of the fabric a similar appearance
Knit: Fabric formed by a continuous thread of interlocking loops; two stitches, knit and purl, are the basis for all knitting constructions
Rib knit: Made by alternating sets of knit and purl stitches in the same row, forming pronounced vertical ridges
Wale (or rib): A column of loops along the length of a fabric

3.10 Clothing Construction Terminology

Dart: Tapering stitched tuck
Hem: Border or edge of cloth made by turning the edge in and sewing down
Interfacing: Semi-stiff material between two layers of fabric in a garment, used in collars, lapels, cuffs, etc.
Lining: Layer of material attached to inside of garment
Pleat: Folds of fabric that provide fullness
Seam: Line of junction between two pieces of fabric that are sewn together

Sleeves:

Kimono: Sleeve cut with the rest of the garment, rather than a separate panel
Raglan: Sleeve that joins the bodice in a diagonal seam extending to the neckline area
Set in: Sleeve that joins the garment in a seam that encircles the arm over the shoulder

Tuck: Slender fold of fabric that can be stitched along all or part of its length
Velcro: Hook and loop tape that intermesh when pressed together

3.11 References

Adolf, F.P., The structure of textiles, in *Forensic Examination of Fibres*, 2nd ed., Roberston, J. and Grieve, M., Eds., Taylor & Francis, London, 1999.
Lin, A.C.-Y., Hsieh, H.-M., Tsai, L.-C., et al., Forensic applications of infrared imaging for the detection and recording of latent evidence, *J. Forensic Sci.*, 52(5), 1148–1150, 2007.
Petraco, N. and Kubic, T., *Color Atlas and Manual for Microscopy for Criminalists, Chemists and Conservators*, CRC Press, Boca Raton, FL, 2003.
Raymond, M.A. and Hall, R.L., An interesting application of infrared reflection photography to blood splash pattern interpretation, *Forensic Sci. Int.*, 31(3), 189–194, 1986.
Ristenbatt, R. and Shaler, R.C., A bloodstain pattern interpretation in a homicide case involving an apparent "stomping," *J. Forensic Sci.*, 40(1), 139–145, 1995.
Robertson, J. and Grieve, M., Eds., *Forensic Examination of Fibres*, 2nd ed., Taylor & Francis, London, 1999.

Stoilovic, M., Detection of semen and blood stains using polilight as a light source, *Forensic Sci. Int.*, 51, 289–296, 1991.

SWGMAT Trace Evidence Recovery Guidelines, *Forensic Sci. Commun.*, October 1(3), 1999.

Taupin, J.M., Clothing damage analysis and the phenomenon of the false sexual assault, *J. Forensic Sci.*, 45(3), 568–572, 2000.

Textile Institute, *Textile Terms and Definitions*, 11th ed., Manchester, England, 2002.

Vandenberg, N. and van Oorschot, R., The use of polilight® in the detection of seminal fluid, saliva and blood stains and comparison with conventional chemical-based screening tests, *J. Forensic Sci.*, 51(2), 361–370, 2006.

Stains and Deposits

4

4.1 Introduction

Stains and deposits on clothing and other surfaces result from people's interactions with each other and their various environments. This includes natural environments as well as human-mediated environments resulting from industrial, agricultural, horticultural, domestic, or recreational activities. Stains occur when a liquid or liquid suspension is absorbed into a substrate material. Deposits result from materials, whether liquid or solid, that adsorb or adhere to a substrate. The term "deposit" will sometimes be used in this chapter to include stains.

If the activities that produced the stains and deposits occurred during the incident under investigation — whether a crime, an insurance claim, or a determination of liability — the stains and deposits can shed light on what transpired, in which locations, by what means, and in what order. A set of activities characterizes every event. Any resulting deposits serve as markers of those activities.

Another set of deposits reflects the normal activities of daily life. Such deposits may be used as reference samples, if the source material or source activity is known, or as secondary standards, if the source is not known but can be inferred. The food people cook, the body fluids they secrete or expel, residues from the plants and animals around them, materials they use to make and repair objects, greases and lubricants from the vehicles they use for transportation, and materials used or produced in their workplaces often transfer to clothing items.

Transferred material can link objects, persons, and events and may be used in the reconstruction of an event. These materials and their residues are encountered as stains, deposits, traces, and debris. Together, the several classes of transferred materials constitute what is usually considered trace evidence. In this chapter, we focus on stains and deposits. Traces and debris will be discussed in Chapter 8.

There are several reasons to examine stains and deposits on clothing:

- To provide information about transfer of materials and manner of deposit
- To provide clues for investigators
- To aid in reconstruction of events
- To decide which materials to test further and which to simply describe
- To alert forensic scientists in other disciplines to evidence they may need to examine
- To alert police and investigators, who may have noticed potential sources of stains and deposits described by the scientist and may be able to obtain samples
- To provide a record for future work if a new suspect, site, or evidence is developed, even many years after the original examination, and often by different examiners.

The potential value of stains and deposits can be realized only if the working scientist is able to base conclusions on objective data and perform examinations under known conditions. The forensic science and materials science literature is replete with information about the identification and source attribution of the materials that produce stains and deposits. Very little has been published regarding the types of information provided by stains and deposits themselves. Objective data for non-numerical observations implies a standard descriptive terminology. The existing literature for bloodstain patterns (see references in Chapter 5) and soil aggregates (Soil Survey Staff, 2006) provides useful descriptive terminology for their deposits on clothing, but terminology for deposits of other materials, such as paint film defects (Hamburg and Morgans, 1979) and plant disease patterns (Plant Disease Patterns, accessed 2008), is not directly applicable. [A suggested standard terminology has been proposed (Cwiklik, 2006) but not yet published; we provide it in Section 4.10.]

CASE EXAMPLE 4.1*

Providing clues

The dismembered body of a young man was found in a large city park, and another dismembered body was found in front of a large apartment complex in the same city. Each body was wrapped in an inexpensive waxed paper of the type used in bakery shelves, then again wrapped in carpet. Debris on the clothing of both victims included whole-wheat flour; deposits included splashes of white bean soup (both were identified using light microscopy following the initial examination with a stereomicroscope). This suggested that the two homicides were related. In the early 1970s, when the crimes occurred, these were not common foodstuffs, but were used in foods sold at bakeries operated by a then-radical ethnic religious group focused on community and self-sufficiency. The waxed bakery paper was the type used in such bakeries. The splashes suggested food that was in the process of being prepared rather than material transferred from the bean pastries and other foods sold there. When forensic scientists suggested such a bakery as a crime site, detectives were skeptical — the group had the reputation of personal discipline and a low crime rate — but they pursued the lead and arrested the brother of a man who worked at one of the bakeries and lived in adjoining living quarters. The brother had stolen the bakery keys to conduct narcotics transactions there without the worker's knowledge. By the time the site was located, the carpet had been replaced, but the nonstandard measurements of the two areas with new carpeting matched the dimensions of the carpets in which the bodies were wrapped. Additional corroboration of the brother's involvement in the murders was provided by body fluid evidence at the bakery and trace evidence in the apartment.

* Information provided by Mary Jarrett-Jackson, formerly with the Detroit Police Department Crime Laboratory.

CASE EXAMPLE 4.2

Establishing a time line; deciding which samples to test

A man was found badly beaten on the floor of the suspect's residence. The victim was taken to a hospital, where he died after his wounds were treated. The suspect admitted to hitting the victim on the head with a clothes rod in an attempt to ward off an assault. The suspect said he hit the victim and removed the victim's pants to restrain his movements. The suspect then carried the clothes rod outside and discarded it. Police found a wooden clothes rod in wet grass outside the residence and submitted it to the laboratory. The laboratory was asked to evaluate whether the clothes rod could have produced the injuries that fractured the skull of the deceased, or whether another weapon, and perhaps other assailants, produced those blows.

Damage to the wooden dowel suggested that it was cracked by being pulled down from the brackets in a closet; several hairs were observed in the fresh break, but only a single droplet of blood and a small area of diluted blood (diluted by the wet grass). The dowel, not sturdy to begin with, was cracked from another event; although it may have been used to hit the victim, it was not likely to have fractured his skull.

A round wooden table found at the scene exhibited extensive blood spatter and linear impact damage to the wood. Several hairs were embedded in blood and apparent hair swipes in the blood suggested contact with a scalp. The linear impacts exhibited a sharp edge, unlike the rounded dowel, and cracked the wood of the table. Therefore, a weapon other than the dowel was used to produce the damage to the table. If the blood spatter and the damage to the table were produced during the same incident, the other weapon may also have been used to bludgeon the victim.

The table was finished with wood veneer. The wood veneer was separated in the damaged area, exposing fresh wood. A tiny droplet of spattered blood was observed on the exposed wood. Therefore, that particular droplet was spattered after the blow that split the wood. A smear of blood was deposited across the dented area, but not across tiny splinters of wood in the line of the smear. Therefore, the smear was produced before the blow that splintered the wood. The blood spatter and the blows were thus part of the same event, indicating that a weapon other than the dowel was used to produce not only the damage to the table, but also the extensive blood spatter on the table. The hairs and apparent hair swipes on the blood provided a link to a person involved in the incident, either to someone with a bloody scalp or to someone whose head touched the bloody table while the blood was wet.

The several hairs that were embedded in blood were collected for comparison with the victim's head hair. Which deposits of blood should be tested? The blood droplet in the fresh wood exposed by a blow and the blood smear that was separated by the blow were both selected for further testing, because they exemplified deposits before and after a blow to the table, thus spanning the sequence of events.

4.2 Information from Preliminary Examination

4.2.1 Overview

Initial observations should begin with a visual examination of the item of clothing, using the naked eye. The examiner should pay particular attention to anything that appears out of place. General categories include deposits of solids, liquids, and polymers manifested as the following:

- Stains and spatter
- Crusts, films, and gummy deposits
- Dust; powdery matter
- Aggregates of particles, lint, or debris, or a change in luster or sheen
- Mechanical damage, such as cuts, tears, punctures, and other holes
- Thermal, chemical, or photochemical damage, indicated by melting, bubbling, film formation, fading, or color change, and biological damage, such as mold growth, moth holes, and so on

(Also see Chapter 6.)

Each stain and deposit should be documented on a detailed diagram of the clothing. The diagram should include some details of the garment construction, such as pockets and buttonholes, and the location and general appearance of stains, deposits, and damage. The direction and sequence of damage and deposits should be noted. Stains and deposits should be observed from both the inside and outside of the garment. Each stain and deposit should be examined with a stereomicroscope to obtain preliminary information. Deposits and damage on an item should be checked against corresponding areas of other clothing reported or known to have been worn by the same person at the same time.

A preliminary microscopic description of each stain and deposit should accompany a detailed diagram in the laboratory bench notes (Figure 4.1). For example, what appears to be blood is described as a "reddish-brown glaze, round spatter droplet, approx. 1 cm." The results of a presumptive chemical test for blood should follow. If the test results for blood are negative, the analyst can ask what other material it might be and proceed with observations that might permit a general classification of the material that would enable the analyst to decide whether to pursue further testing. For example, if a deposit initially suspected of being blood is examined with a stereomicroscope, the data might consist of observations such as oily stain margins with very fine particulate solid deposit. This would suffice to rule out blood, which would be a translucent or transparent glaze, and might suggest the deposit was a wood stain.

The microscopic observation is a preliminary analysis, much like the presumptive chemical tests for blood and semen, and can be objective and rigorous if it is based on specific data. Data and results from stereomicroscopic observation may include descriptions such as: "Apparent tan soil, smeared damp — heterogeneous caked deposit of mineral grains and plant parts"; "Red-brown glaze, on surfaces of threads only," describing a light contact of apparent blood; "Looks like honey or tree sap — clear amber-colored viscous glaze, sticky when touched with probe"; "Plant part — seed husk???" The latter description should be accompanied by a sketch. The question marks indicate that a possibility is being

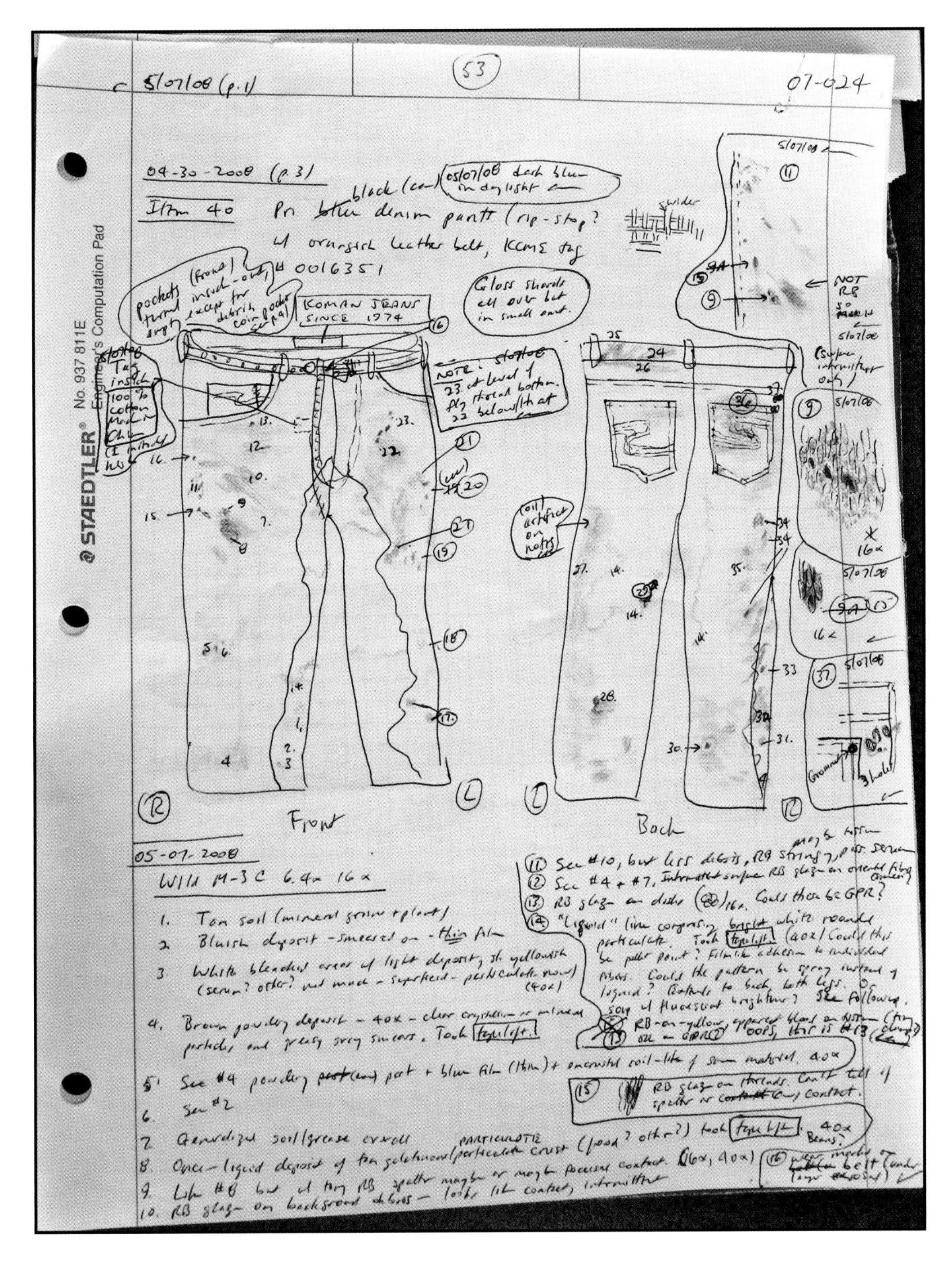

Figure 4.1 Diagram of shooting victim's blue jeans and descriptions of stains and deposits. Prior activity was of interest.

suggested but is not confirmed or even narrowed. Should it later be of interest, further examination or testing would be required to confirm or refute the preliminary hypothesis. Small sketches of individual deposits are often useful, especially for complex structures. The magnification(s) should be recorded. (Tools for classifying deposits are described in Appendix 4.)

The following sections discuss what can be accomplished by a preliminary stereomicroscopic examination. What is it? Where did it come from (material or origin)? How was it deposited? How did it transfer to the evidence item? What happened to it afterwards? When was it deposited? Is it from this event or from something else?

4.2.2 Class of Material

It is useful to be able to recognize common types of materials at the magnification levels provided by a stereomicroscope, even if the specific material cannot be identified without further examination or testing. Examples of such materials include plant parts, body fluids, paint and coatings, hairs and fibers, cosmetics, foods, and architectural or building materials (Petraco and Kubic, 2003). Plant parts may be encountered in processed form, such as paper, natural fabrics, and treated wood. Paint and coatings include nail polish, whitewash, and material painted on trees to discourage insect activity. Oil stains may be from food, automotive sources, lubrication, or hair products, spanning classes of materials. Some of these materials are discussed in more detail in Chapter 8. The focus of this chapter is the stains and deposits themselves.

Reference collections comprising various materials found in our environment are indispensable. Such collections may begin with reference materials for individual cases and can be developed as training topics or sampling opportunities arise. We provide a list of suggested materials to include in such a collection in Appendix 3.

The description of a deposit should include specific observations as data and should provide a mental image for the reader. For example, comments appended to the data, such as "looks like jam," are useful even if the material later proves to be something else. The informal "looks like" indicates a comment rather than a conclusion. When a final identification or classification turns out to be different from the preliminary comment, this should be cross-referenced in the laboratory notes.

4.2.3 Appearance of Deposit

Each deposit has a morphology that can be described, such as particulate residue, a crust, a film, a glaze, a caked deposit, or a drop. This morphology can be directly related to how the deposit was transferred to the substrate. For example, kneeling in dry soil might result in a heterogeneous particulate deposit on the knees, but kneeling in mud might result in a caked deposit, as might a mud splash. A splash from a puddle in the road would result in particulate residue of a liquid suspension (Figure 4.2), recognized by the typical rounded shape of a liquid deposit even if no liquid remains.

4.2.4 Manner of Deposit

It is usually possible to determine whether a material has been smeared, splashed, spattered, or dripped, or whether it was deposited by light or heavy contact. Soil, blood, or nail

Figure 4.2 Particulate residue from splash of road dirt.

polish could be deposited while wet or tacky, or rubbed via more forcible contact with an already-dry deposit. A deposit of material could originate from the outside of the garment or from the inside. The direction of a smear can often be determined. Particulate deposits can be scattered or broken into chips or shards when force is applied, and they may be deposited by impact (i.e., when either the deposit or the substrate is smashed, crumbled, melted, or otherwise deformed by the act of transfer). For example, a soil-encrusted stone may scatter soil, or a bullet skimming the surface of a caked food deposit may smash it or smear it. Liquids can splash or create secondary spatter from impact (Figure 4.3).

The manner of deposit provides an additional level of information to that obtained from identifying the class of material or appearance of deposit. A hierarchy of propositions to use in forming and comparing hypotheses includes three levels of information that may develop from casework examinations: source, activity, and offense (Cook et al., 1998). Source-level information includes identification or classification of a material or body tissue and correlation with a source object, person, or animal. Examples include obtaining a DNA profile from an evidence blood sample and correlating it with an individual, or identifying a fiber as Nylon 6 and correlating it with a carpet at a crime scene. Activity-level information addresses action and its effects, and can be used to reach conclusions about what happened. Examples include the aforementioned blood sample being spattered, or the nylon fiber being wedged in a wooden plank that was used to bludgeon a victim. Class of material and appearance of deposit address the source level. Manner of deposit addresses the activity level.

Offense-level information addresses the crime or regulatory violation that was committed. Examples include self-defense versus murder, suicide versus homicide, homicide versus disposing of a body. A single stain or deposit can seldom provide information about an offense, but it might be used in a reconstruction of events that does address that level of information. Reconstruction often rests upon activity-level information, as illustrated in Case Example 4.6.

Figure 4.3 Mud scattered on shirt by mud-coated stone thrown at a person. Note relative absence of mud in center and spatter at margins.

4.2.5 Sequence of Deposit and Time of Deposit

Stains and deposits can provide a timeline, especially when several related deposits can be evaluated with respect to one another. For example, fibers and particles embedded in wet or tacky paint or blood must have been deposited before the substrate material dried. Accumulations of material and accumulated wear can provide a relative measure of the passage of time. Accumulations of debris and wear can assist with determining whether damage, for example, is fresh or old. This is further discussed in Chapter 8.

Sequential deposits, such as the appearance of one stain or deposit on top of another, provides information about the sequence of events. Another example is damage occurring either before or after a deposit. This is illustrated in Case Example 4.2.

4.2.6 Deposit from the Outside or the Inside Surface

Solid materials can be deposited on either the outer or inner surface of a garment. This is also true of liquid deposits and stains. However, liquids can soak into a fabric and appear as stains on both the inside and the outside surfaces. It is possible in most cases to determine from which side a liquid was deposited, even if it soaked through. One indicator is the extent of the stain; the area of the stain or deposit may be more extensive on the side from which it originated. Another indicator is the staining of surface fibers, which are typically stained on the side on which they were deposited but not always on the opposite side. A third indicator is wicking action, a chromatographic effect in which components of a stain or liquid deposit begin to separate as the liquid spreads. This is often more prominent on the surface from which the stain originated. A related indicator is a thick edge of deposit at the margins of the stain, also usually more prominent on the surface from which the stain originated.

Oily deposits are more difficult to interpret, in part because the oil has seldom completely evaporated. This precludes the transport of residues and the formation of a distinct outer margin.

4.2.7 Direct or Indirect Transfer

A material can be deposited onto a garment directly from a primary source or as a secondary transfer from the item on which it was first deposited, referred to as the primary substrate. Blood from a nosebleed that is ejected and spattered onto the shirt of the person who is bleeding exemplifies primary transfer. If this shirt is removed while the blood is still wet and placed in a laundry basket, some of the blood may transfer as an indirect (secondary) deposit to another item; for instance, to another garment in the laundry. Some of the blood may also transfer from the shirt of the person who was bleeding to the shirt of another person who was trying to stop the bleeding. If the bloody garment is firmly pressed against another fabric item, a careful examination may be required to determine that the deposits on the recipient item are indeed secondary transfer.

A liquid or suspension such as blood could initially be deposited from the outside of the garment or from the inside (for example, if the shirt is rolled up or if the skin beneath it was cut). It could also transfer from a different location on the same garment through a fold (secondary, tertiary, or higher degree of transfer). The deposit could be complex: the liquid could have soaked through the garment from the outside, pooled on the skin, then reabsorbed to the same or another location from the inside (Figure 4.4).

Similar considerations apply to transfer of particulate deposits, except that particulates are rarely carried through the fabric unless transported by a liquid or forced through by air or another gas. The fine particles carried through a vacuum filter exemplify the latter. Gunpowder residues from contact or near-contact shots that are embedded in the target fabric — and often in a fabric layer beneath — are another example.

Prints (patterned marks) and impressions on clothing are examples of direct (primary) transfer of a pattern via indirect (secondary) transfer of materials. The pattern of shoe prints, tire marks, and fabric impressions found on clothing items can be correlated with a particular shoe, tire, or fabric. The material transferred from the shoe, tire, or fabric may assist with reconstruction of events. As an example, in order to leave a bloody or powdery shoe print on a garment, a person needs to have first stepped in blood or in a powdery material. Examples of powdery materials are fine soil, plaster dust, flour, and soot. Impressions, which are produced by a patterned object pressing against another surface, leave an indented pattern and may or may not include detectable transfer of material. This is further discussed in Chapter 5.

Prints may also be produced when material is removed by contact with a patterned surface. This type of print is rarely encountered on clothing items, unless a suitable deposit on the garment allows a clean enough removal of material to produce the pattern of the contacting surface.

4.2.8 Alteration

The moment a garment leaves the assembly line or the hands of a seamstress or tailor, the process of alteration begins. Alterations subsequent to the purchase and first use of a

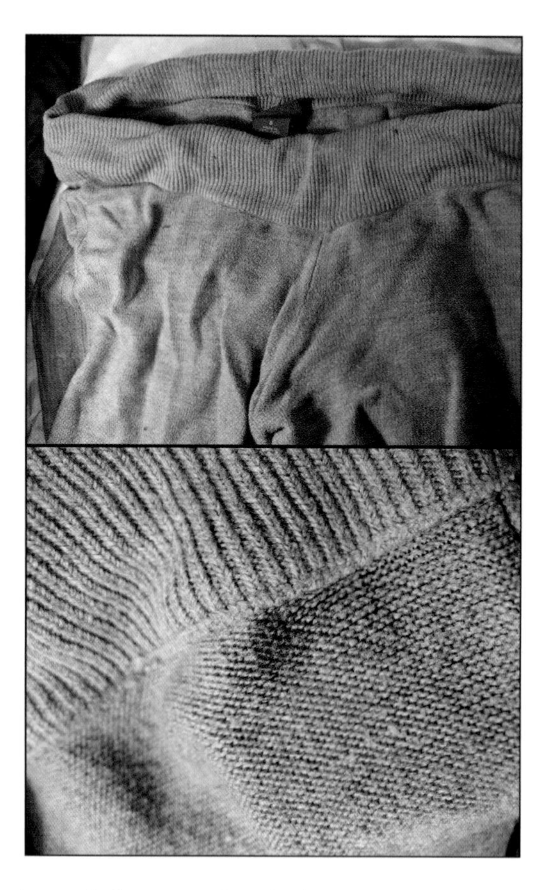

Figure 4.4 (Top) Bloodstain below waistband, deposited from the inside. (Bottom) Deposit viewed from the inside.

garment are usually of most interest in casework. Not only the substrate material, but also stains and deposits, are subject to alteration.

Mechanical alteration to fabrics includes fracture, deformation, stretching, and abrasion. Chemical alteration may produce bleaching and polymer damage, resulting in brittle areas of fabric or congealed areas of partly dissolved material. Bleaching may be partial, which results in diminished color, such as rose-pink bleached areas on dark fabric (Figure 4.5). Other chemical changes include fading, bleaching, discoloration, weakening of fabric resulting from changes to the polymeric structure, and hardening of some deposits. Some materials react over time with oxygen or light, resulting in cross-linking of polymers; this accounts for the difficulty of removing certain older stains and deposits during laundering. Areas of a garment that are cleaner or less faded than the surrounding fabric suggest loss or change in position of an overlapping, adhering, or attached material.

Thermal alteration of fibers may include melting, which deposits a film on the surrounding fabric. Thermal decomposition may result in yellow or brown discoloration, charring, bubbling of a melt, and swelling (Figure 4.6). Friction-generated heat may produce partial fiber melt on synthetic fabrics. This effect may be observed when a pedestrian who has been struck by a vehicle slides across the vehicle hood. The heat generated by a spinning bullet may produce a similar effect around a bullet hole. Biological activity may result in stains, growths, and discoloration from mold and other microorganisms. Other

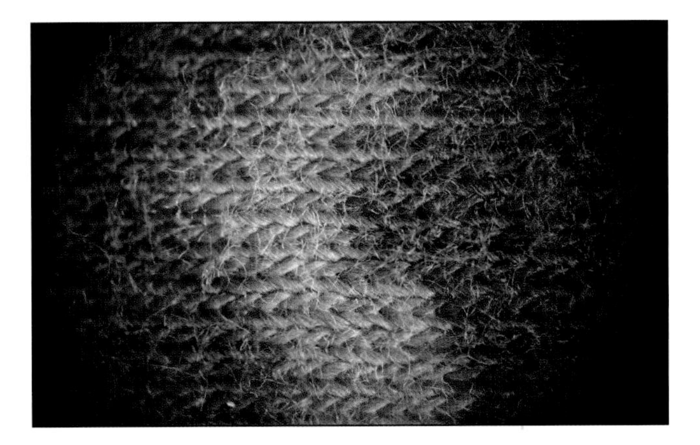

Figure 4.5 Bleach stain on olive-colored knit shirt.

biological activity may include invasion by plant roots, bites by insects such as moths, and scavenger activity, including bites and tears.

Alteration affects not only the substrate fabric, but also materials deposited on the fabric. Mechanical alteration to deposits includes loss from flaking or friction-induced removal of material. An existing deposit may crumble or fracture, producing powdery residues, flakes, and chips. It may be scratched by abrasion or softened by moisture or solvents. A stain may fade or harden or change color. Loss of adhering debris and evaporation of volatile liquids may also occur. Loss of debris will be discussed in Chapter 8.

When petroleum products such as gasoline evaporate, higher-boiling residues may remain, producing faint, slightly translucent stains. Dissolved solid petroleum residues may

Figure 4.6 Melting with decomposition during experiment with burning blue jeans. Melt is from plastic button shanks and plastic thread around zipper.

leave trace residues when the liquid portion evaporates (Mann,* personal communication). Generally, translucent or nearly translucent stains result from a refractive index effect. The substrate fibers are essentially mounted in the liquid or in a transparent or translucent solid or wax. This applies to oily stains and to residues of petroleum products. If there is little material on the fibers, the area may be observed as simply having different luster.

Finally, microorganisms or insects may consume blood and other food deposits, diminishing the deposit, often leaving only an outline of the original deposit. The type of mold, fungal, or bacterial growth observed on a deposit can sometimes indicate the nature of the original material if none remains. This may be of particular interest if a bloodstain or other body fluid or tissue deposit is suspected, and it would require examination by a microbiologist. This is illustrated in Case Example 4.3 (McIntyre,** personal communication).

CASE EXAMPLE 4.3**

Moldy stain — was it once blood?

Police pursuing leads in a homicide found a large stain in the trunk of a suspected transport vehicle and wanted to know if it was blood. When it was examined in the laboratory, preliminary tests for blood were negative, but the stain had largely been consumed by fungal growth. The stain was cultured and the culture examined. It was found to be a type of growth that would not be supported by blood. This allowed police to pursue other leads instead.

** Information provided by Lynn McIntyre, Washington State Patrol Crime Laboratory, Seattle.

4.2.9 Wear

Wear can be defined as the accumulation of small alterations and damage resulting from repeated contact. The garment itself may exhibit wear. For example, surface-dyed fibers in blue denim pants may be worn away by repeated abrasion to expose white or lighter-colored fabric beneath (Figure 4.7). At first glance, the worn area may appear to be a stain or deposit, but this is easily distinguished from a stained or bleached area under a stereomicroscope. An abrasion is evidenced by roughly severed fibers protruding from the threads and aligned in one direction. Bleaching removes color but does not involve mechanical damage (Figure 4.8). Wear may be accompanied by accumulated deposits, such as those observed on the knees of a pair of work pants that are in frequent contact with greasy concrete. Both repeated abrasion and accumulated grease and soil deposits are observed. Such deposits cannot usually be associated with a single incident, but may provide information about activities that are characteristic of the wearer. Accumulated deposits may transfer, providing links between the wearer and other objects or persons.

When the deposits are subjected to wear, crusts or films may crack, eventually producing tiny flakes. Scratches and pits may develop. Stains may fade. Debris may accumulate on oily and waxy deposits as well as on softer paints and other films. Debris may also be deposited on fracture surfaces and at damage margins. Evidence of wear should be described in laboratory notes; for example, "gritty white opaque crust — not recent — cracked and flaked."

* Mann, D.C., MDE Forensic Laboratories, Seattle, WA.
** McIntyre, L.D., Washington State Patrol Crime Laboratory, Seattle, WA.

Figure 4.7 Abrasion on blue denim, resulting in exposed white fibers and rupture of surface threads, now upright along the thread margins. (25× magnification)

If a garment was collected or otherwise removed from use shortly after the incident under investigation, wear on a stain or deposit would suggest that it is probably unrelated to the incident. However, it may still be useful in establishing a history of activity. In cases of suspected sexual crimes, the scientist may wish to analyze older deposits of semen to determine

Figure 4.8 Bleach stain on blue denim, discoloring the surface fibers, which remain in the normal position on the threads. (40× magnification)

ongoing sexual assault or abuse, or sexual activity with several partners. It may be useful to conduct a few experiments to establish approximately how much wear will produce the observed damage. For example, test results might indicate that a crust of plaster or grout does not survive a single wearing without cracks. The criteria for prolonged wear might then be an accumulation of fibers and particles at the fracture margins.

4.2.10 Alteration from Immersion in Water and Alteration from Burning

Extensive damage to clothing that has been submerged in water or exposed to flame or high heat is an expected effect. Deposits on the clothing are similarly affected. Such garments should nonetheless be examined for stains, deposits, and even debris. It is surprising how much information sometimes remains, especially in protected areas such as seams, pockets, and folds. The garments are often lined with skin and hair, which should be collected, at least in part, to serve as secondary reference samples. Stains and deposits should be evaluated for what they might have been before they were damaged, requiring some knowledge of how materials are affected by heat or by exposure to water and the organisms in it.

CASE EXAMPLES 4.4

Finding evidence after garment was washed

A: Probably washed by hand; some bloodstains surviving*

A young man involved in a fight with several other men was cut in the throat, held from behind while blood flowed. A leather jacket worn by one of the potential assailants was examined for blood, but none was observed upon initial examination. The potential assailant told police that he had washed the jacket. The scientist, reasoning that the jacket was probably wiped clean rather than laundered in a washing machine, looked in the seams and crevices using a stereomicroscope and found blood on the sewing threads in the seams of the arms. This would be the likely location if the wearer was the person restraining the victim. No stains or deposits were observed on the leather surfaces; not surprising had the visible blood been washed off.

B: Probably washed in a washing machine; some bloodstains surviving**

A woman was raped and beaten, resulting in extensive bloodshed. A suspect arrested the next day had apparently discarded his outer clothes, but police collected his tennis shoes and submitted them to the laboratory. The shoes were bright white and appeared to have been washed. There was no visible blood, not even on the laces, which might have retained some residues. The scientist who examined the shoes noticed that the laces were tan instead of white. Sales personnel at a local shoe store verified that the shoes would be sold with white laces. Thinking there might be some traces of blood remaining, the scientist examined all the fabric surfaces and threads with a stereomicroscope and found reddish-brown stains on the threads at the side soles. The stains were found to be blood and corresponded with the victim.

* Information provided by Kerstin Gleim, Emerald City Forensics (formerly with the Washington State Patrol Crime Laboratory).

** Information provided by George Chan, Chan & Associates (formerly with the Washington State Patrol Crime Laboratory).

4.2.11 Alteration from the Examination

Every time an item of evidence is removed from its container and placed on the examining surface, some alteration and loss take place. Loosely adhering materials fall off onto the paper placed on the examination surface under the item or are transferred to other areas of the garment. Fragile deposits may crack and flake off or crumble into dust. This is unavoidable, but it can be minimized by careful handling. The practice of scraping a garment to collect trace evidence should be used only when there is no good alternative (for example, when the entire garment is heavily encrusted with mud and plant debris), because of the potential information loss from disruption of deposits. However, under other circumstances — for example, when searching for shards of glass — gently shaking the evidence or scraping it would be strongly recommended. If employed, scraping should be done only after other examinations have been performed and less-disruptive methods used. This is discussed in Chapter 8.

Swabbing and chemical mapping using overlays also alter the evidence, but their value usually outweighs the need to avoid alteration. It is helpful to recognize these methods as sampling techniques for the test that follows. Before sampling a particular area, it is important to examine the garment surface with a stereomicroscope. Sampling includes wet and dry swabbing and mapping for semen, saliva, or gunshot residue via application of filter paper or photographic paper. Any method alters the item being examined if it results in the removal of debris, the softening or partial dissolution and spreading out of water-soluble deposits and stains, the microscopic movement of particles and fibers, or microscopic abrasion of deposits. Wet swabbing of body fluid stains can alter their appearance enough to result in significant loss of information about how the material was deposited. It is critical to complete a thorough examination and documentation of the area to be sampled prior to sample collection. This includes sampling for preliminary testing, such as swabbing and tape lifting.

Thin deposits of blood are notoriously difficult to see on black surfaces, but they can usually be detected as deposits of reddish-brown glaze via examination using a stereomicroscope. Examiners of biological evidence should conduct and document microscopic examinations prior to removing or disturbing stains and deposits for presumptive testing. In some instances, particularly when a deposit is very small, a chemical presumptive test would consume the sample that might be better reserved for DNA testing. In that situation, a microscopic examination might substitute for chemical testing. Finally, the practice of swabbing entire areas for possible blood may result in combining DNA from more than one deposit on the same swab. This can be avoided if microscopic deposits are located using a stereomicroscope and if stains are located with an alternate light source.

Collecting traces and debris on tape lifts or other lifts such as sticky-notes also alters the evidence. The material removed can no longer be evaluated for manner of deposit or for co-deposit with other particles and fibers. A superior method for collecting fibers is to pick them manually with forceps, and for both fibers and particles to lift them with a sharpened tungsten probe. When used judiciously after manual collection, tape lifts and sticky-note lifts can be a valuable method for collecting traces and debris. This is discussed again in Section 4.4.2 and also in Chapter 8.

4.2.12 Relationship to Other Stains, Deposits, or Damage

No piece of evidence exists in a vacuum. The relationship of deposits and damage to other evidence on the same or associated garments should always be considered. For example,

Figure 4.9a Contact deposit of hot chocolate on sleeve cuff, transferred up the sleeve when the shirt was removed.

stains may soak through folds to other parts of a garment, to other garments, or to bedding, furniture, car seats, or flooring. A stain or deposit may transfer its near-mirror-image to an area of fabric with which it is in contact (Figure 4.9). Transfer may continue to occur after the incident under investigation; for example, as a victim continues to move or as clothing is moved and removed during rescue efforts. It is thus incumbent upon the scientist to examine any photographs depicting the clothing, at the scene before the person or body was moved and at a subsequent location. Fluids or debris extrinsic to the incident under investigation, such as pooled blood or automotive fluid at an accident scene, can also transfer to clothing, potentially producing stains unrelated to the actual incident. Stains and deposits can be transferred during packaging of the item too. This can sometimes be referenced to stains and deposits on the packaging itself. Bearing in mind these caveats, stains that are soaked through a garment may provide links with a scene or other items.

Stains that are continuous from one item to another can assist in positioning items relative to each other. For example, was a shirt inside or outside a pair of pants, or were two items in contact during simultaneous use, such as a pajama top and a bed sheet? Some deposits reflect the motion used to produce them. If damage exists parallel to a deposit, such as a linear abrasion parallel to a paint smear, one might infer contemporaneous production of the deposit and the damage; that is, the damage happened at the same time as the smear. Whether a garment has been worn inside-out can often be determined by a correlation of stains and deposits; if not, it can usually be established by transfers of traces

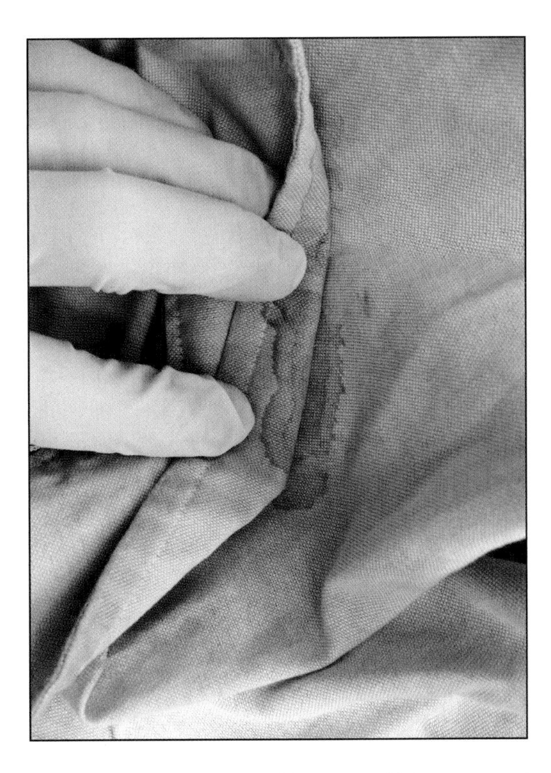

Figure 4.9b (Continued) Near mirror-image of lower cuff deposit and transferred deposit.

and debris (Chapter 8). Finally, a correlation of stains and deposits can provide a basis for logical sample selection for further testing.

CASE EXAMPLE 4.5

Identification of material class, manner of deposit, and corresponding deposits

An important question in an investigation was whether a man was unfaithful to his wife while she was away on a business trip. When the wife returned, she reported finding what looked like lipstick smears on a shirt in the laundry basket. It was a different shade than hers, and she stated that the shirt in question had just been washed before she left. Light smears of a soft, waxy reddish material that appeared to be lipstick were found on the front of the shirt on the button placket as well as on the corresponding area of the buttonhole placket. Reference samples of lipstick exhibited similar composition when examined with a stereomicroscope; this was confirmed with polarized light microscopy. The deposit on the button placket, covered when the buttons are fastened, could not have been deposited while the shirt was buttoned, unless the lipstick or something bearing a deposit of lipstick had been inserted under the buttonhole placket and smeared onto the button placket only. This suggested that the lipstick was deposited while the shirt was in the process of being unbuttoned. While not proof, this finding lent credence to the wife's suspicions and provided a lead for investigators.

CASE EXAMPLE 4.6

Manner and direction of deposit; offense-level information

The body of a man found on a busy roadway had been run over by a vehicle. The clothing was submitted to the laboratory to search for any paint deposits that could provide information about a hit-and-run vehicle. Automotive paint smears were found on the shoulder of the victim's jacket, a location where paint is often found when a person is struck by a car and flipped over the hood. However, the paint smears were deposited in the opposite direction from what would be expected had he slid across the hood; the smears were observed going up the shoulder instead of down. This laboratory finding led police to investigate further. Subsequent investigation established that the victim was killed, then left in the roadway and run over to draw attention away from murder. Comparison of paint reference standards suggested a make, model, and range of years of the source vehicle (Cwiklik and Gleim, 2009).

CASE EXAMPLE 4.7*

Stain pattern on clothing — correspondence with scene

A woman was killed and her body was found outside a 400-unit apartment building — a bloody blanket was found in the apartment dumpster. A man who had lived in the building became a suspect after coming to police attention for trying to set a fire in the dumpster. The victim was linked to the blanket via trace evidence on her clothing, but only a weak link existed between the blanket debris and the suspect's apartment — he had moved out shortly after the murder, selling all his furniture to various neighbors. The building manager subsequently cleaned the man's apartment, then rented it to someone else. Police collected fiber reference standards from the carpet and furniture items sold to neighbors, and laboratory examinations established that the furniture fibers corresponded with predominant debris fibers on the blanket and the victim's clothing. This provided a strong but not definitive link, so police obtained a search warrant to examine the apartment again, this time requesting the assistance of the forensic scientist. They found bloodstains that were soaked into the wood flooring beneath a carpet and blood residues on the underside of the carpet itself. The carpet and floor had been cleaned, confounding attempts to group the blood. The stains on the flooring and carpet were photographed and also traced onto a large piece of thin paper, then compared with tracings of bloodstains on the blanket. Not all of the blood had soaked through and survived cleaning, but the pattern of the denser and lighter bloodstains on the blanket corresponded with the stains on the floor and the underside of the carpet, as did the shapes of the denser stains. This provided the definitive link to the murder site (Cwiklik, 1999).

* Information provided by Lynn McIntyre, Washington State Patrol Crime Laboratory, Seattle.

CASE EXAMPLE 4.8

Identification of material, manner of deposit, and corresponding deposits

A soldier, recently returned from the war in Iraq, had an argument with his wife as she was preparing to bathe and admitted to pushing her head into the water. He claimed to have come to his senses and pulled her back out, then placed her on their bed to resuscitate her. He said she vomited and then apparently choked. She was dead when the aid crew arrived. Investigators wanted to know whether any vomitous material had indeed been deposited on her or on the bedding. Scene photographs depicted the deceased lying on her back in bed, wearing a gray sweatshirt or fleece and covered with a quilt coverlet and white bed sheet up to her neck. The coverlet was no longer available, but the bed sheet was examined. The pattern of livor mortis depicted in autopsy photographs reflected the position depicted in scene photographs, but it would have been essentially the same had the victim been placed there not long after she died. A white froth was depicted in and around the nose and mouth, oral cavity, and between the lips. Crusty material of food origin found on the victim's sweatshirt and on the white flat bed sheet (the top sheet) corresponded with the stomach contents of the victim; the crusty material was determined to be a direct deposit, not a contact transfer, and was not smeared, as would be expected had the sheet been placed over her after she vomited. This evidence supported the soldier's account that the victim vomited in bed, when she was still alive, and died only afterward.

4.3 Getting Started: Workflow for Examination of Stains and Deposits

After deciding which items to examine first, prepare the work area with a smooth hard paper like butcher paper that does not tear easily. In addition to preventing contamination, the hard paper will collect debris that may fall off the items. Substrates such as plastic-lined hospital sheets that are soft but not smooth are not suitable, because they may not release debris adequately. Good lighting is critical. If the work area is next to a large window, sunlight may provide good lighting during daylight hours. Otherwise, use a strong light directly focused on the work area. Overhead fluorescent lighting is rarely adequate for clothing examination. A stereomicroscope of good quality should be readily available, and it is helpful to have one mounted on a boom stand or a rolling stand when examining large items that cannot easily be placed on a microscope stage. (See Appendix 2 for guidance on selection of a stereomicroscope.)

After removing the garment from its packaging, spread it out on the prepared paper. Observation begins with taking a good look, forming overall impressions of the garment and its condition, and inspecting for damage, deposits, and anything else that may be out of the ordinary. The initial overall observation should form the basis for objective observations described in terms of specific data. Document the initial observations and overall impressions, then sketch a detailed diagram of the item. A diagram need not be artistic to be useful. Its primary purpose is to provide an objective record of the data of observations. An important secondary purpose is to stimulate the examiner to notice each detail and its relationship to the clothing as a whole. Some examiners take photographs at this stage. Not only do photographs provide a valuable record, but streaks or smears that are not readily

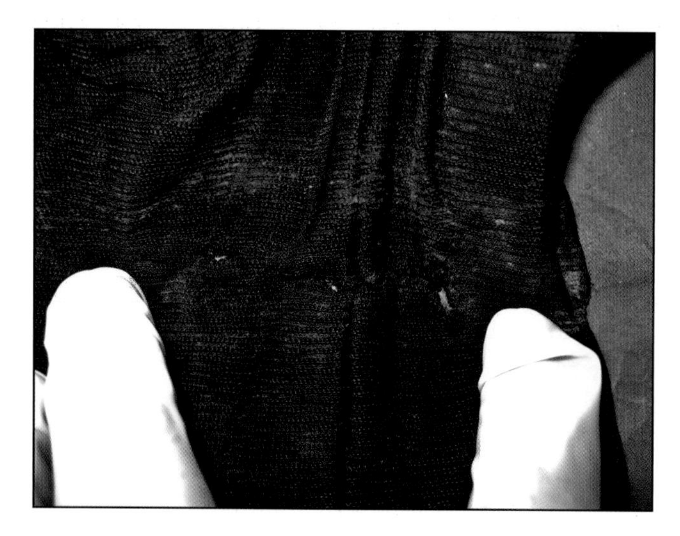

Figure 4.10 Apparent lead fragments at margins of complex bullet entrance.

visible to the eye may sometimes be depicted clearly in photographs. Photographs complement diagrams, but neither should substitute for the other. Unlike overall photographs, a diagram allows the examiner to sort through the profusion of information and record not only focused observations, but also subsequent evaluation.

It is useful to draw enlarged sketches of small areas of particular interest. For example, a lead particle or smear or a lead wipe from primer residue around a fabric defect of uncertain origin might be emphasized, because the finding of lead particles or a lead wipe assists in identifying a hole as resulting from entrance of a bullet or bullet fragment (Figure 4.10). Modern digital cameras simplify evidence photography, so they may frequently be used to supplement, but not replace, a sketch. A photograph may not show the manner of attachment as clearly as a sketch would (Figure 4.11). A summary of what to include in the laboratory bench notes is provided in Appendix 1.

The examiner should document and describe on the diagram every visible stain, deposit, discoloration, abrasion, or damaged area. If there are many deposits, it is useful to number them. This may sound daunting, but after some practice, examiners become efficient as they develop a shorthand for descriptions and incorporate a logical sorting process for what merits more detailed examination. Some of the examinations may be cursory and the descriptions brief (e.g., "yellow-orange crust with plant cells, looks like food, no further exam"), but all should be accounted for in the laboratory notes. If several deposits have similar characteristics, one can simply reference previously described stains, noting small differences (e.g., "see deposit #4, but thicker").

Finally, the examiner should be aware of the case scenario in order to ask what might be expected of the evidence; if, for example, a suspect was really at the scene when certain events took place. The clothing diagram should include notations about areas lacking deposits where some might be expected. These areas should be examined at the highest magnification range of the stereomicroscope in case the deposits are faint, very small, or trace. If the expected material is still not observed, the absence of evidence may disprove a hypothesis. This presumes that the magnification used is sufficient to detect a deposit.

As an example, if the shirt and upper pants of a beating victim exhibit spattered blood, it may be important to know whether the pants were open or closed during the beating.

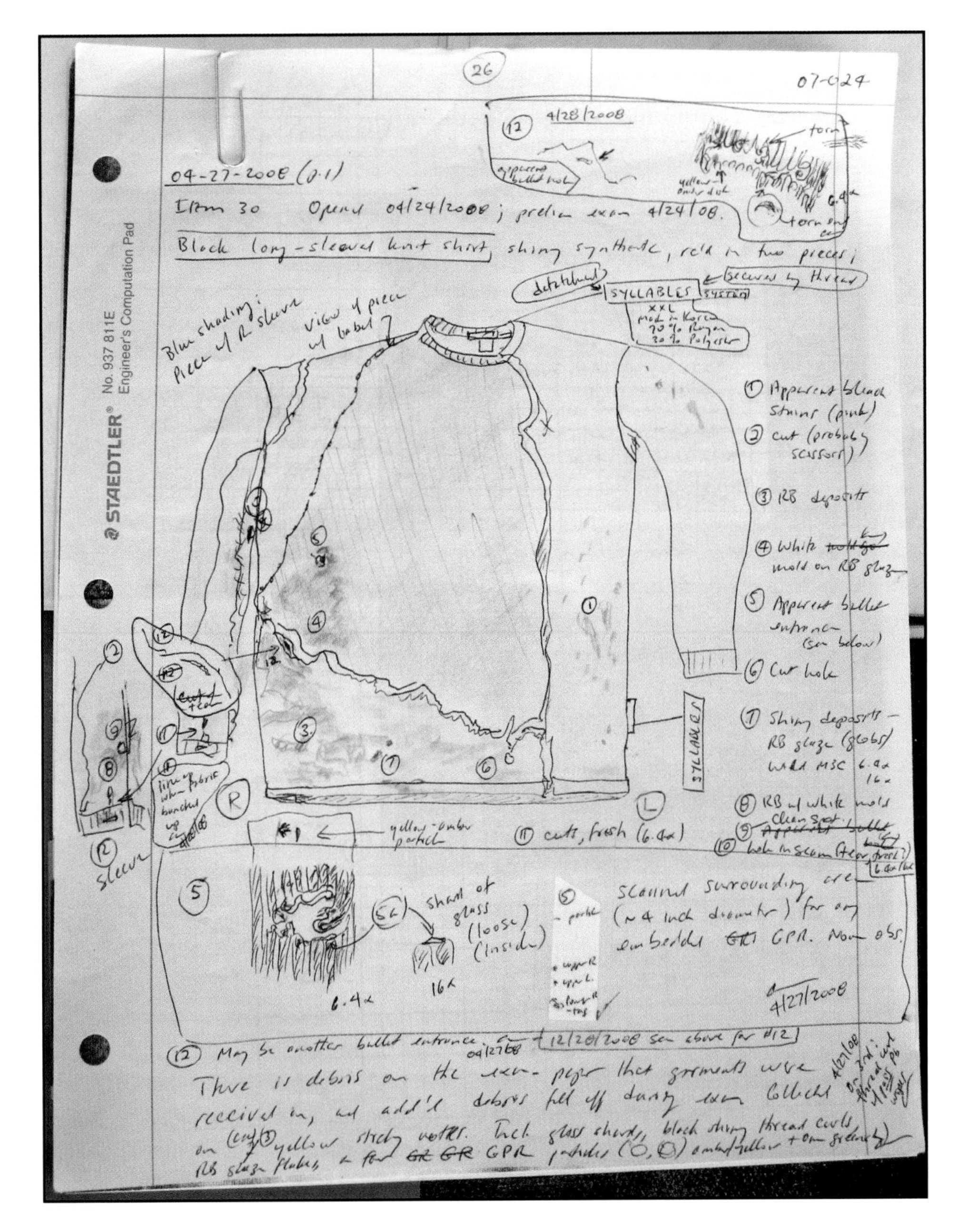

Figure 4.11 Diagram of shooting victim's shirt, with sketch of bullet entry damage.

If no blood spatter is detected inside the fly area, even at the highest magnification of a good stereomicroscope, this indicates that the pants were closed. However, if paint spray were at issue, higher magnification may be required. It may be necessary to pursue examinations with more powerful equipment or use a different methodology. For example, a high-magnification reflected light microscope might be used for fine sparse deposits or an alternate light source used to detect various types of stains. It can also be helpful to sample a small area with a piece of thin clear sticky-tape and examine the sample on the tape using a stereomicroscope. Sometimes a deposit that is barely visible while on the fabric can be observed easily on the tape.

CASE EXAMPLE 4.9

Expected material not found — high-magnification reflected light microscope used

A woman was murdered in her house on a day that several of the rooms were being painted. She was strangled. Her clothing and body exhibited fine paint spray droplets, apparently from decorative paint that had been applied that day by spraying and must have been briefly suspended in the air. Detectives had strong reasons to suspect a man who had access to the house, but they did not have proof. The man freely admitted having been in the house, but he denied being there on the day of the murder. No foreign hairs or other human biological evidence that might provide a link with her assailant were found on the victim. The suspect's clothing contained no paint smears and no obvious paint residues, surprising had he been at the house that day, considering that the painted walls were still wet. The suspect's clothing was examined again. Because of the fine paint spray observed on the victim's clothing, the suspect's belt was examined using a high-magnification reflected light microscope. Very fine droplets of paint spray were observed that corresponded with each of the three decorative colors of paint that were used. This strongly suggested his presence at the time of the murder.*

* Because the original case reference was not located, some of the background information presented is an educated guess.

CASE EXAMPLE 4.10

Expected material not found — alternate light source used

Two men were coming home from a tavern, driving a van despite being heavily inebriated. When the van crashed into a retaining wall, both the driver's and the passenger's airbags were deployed. The van belonged to the employer of the wife of one of the men. Both men exited the van and called the wife to whom the van was assigned. She called her employer and the employer called police. The men had been taking turns driving and neither one could remember who was driving during the impact. The vehicle was examined by investigators, who looked for areas where hair and fiber deposits might have been impacted, but no evidence of who was driving was noted. They submitted the airbags to the laboratory. No visible stains or deposits that could be associated with either man's position in the vehicle were observed. When the items were examined with an ultraviolet lamp, tiny fluorescent spatter droplets were observed, probably saliva ejected with breath expelled upon airbag impact. Material in the saliva may fluoresce. In this case the saliva would have been mixed with beer, which reliably produced fluorescent stains in simulation tests and retained fluorescence three months later (the time elapsed between the incident and the observation under UV light). The individual droplets were each circled and the items forwarded for DNA testing to determine which person was behind the wheel.

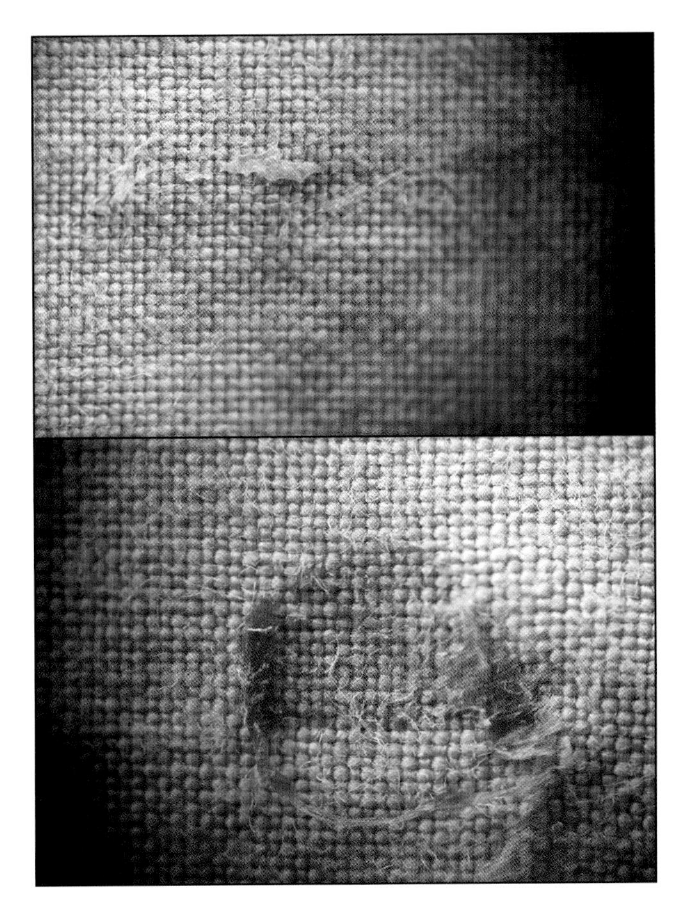

Figure 4.12 (Top) Apparent brown stain on blue cotton pants, with colorless adhering curl of plastic film. (Bottom) Reverse shows a coating on the cotton fabric has melted with decomposition, peeling away and soaking through to the outside.

4.3.1 Examining Individual Stains and Deposits

Regardless of the initial focus of examination, the entire garment should be examined. Examine both the front and the back, and from the inside and the outside. Each stain and deposit should be examined and documented. Examine stains from both "sides" to determine if they were originally deposited from the inside or the outside. Not all solid deposits need be examined from both sides, because solids do not soak through. However, deposits of uncertain origin, deposits of potential significance in reconstructing events, and deposits that may be associated with impact or damage should be examined from both sides. So should all contemporaneous fabric defects (holes and distortions). A deposit on one side may result from or may accompany damage to the other side (Figure 4.12).

4.3.2 Smears and Directional Contact Deposits

When smears, scratches, or other directional deposits or marks are observed, the scientist should determine the direction of the smear or mark if clear and sufficient data exist. Material is often pushed up against the threads and accumulates there preferentially, or is pushed up against itself, in either case revealing the direction of deposit. Examples of materials that exhibit this behavior include caked and powdery particulate deposits such as food and cosmetics, clay and soil, paint and caulking, crayons and waxes, and blood and other proteinous materials (e.g., semen, food, and some types of glue). Surface fibers often align in the direction of a smear (Figure 4.13). Surface fibers in the surrounding area should be used as a reference point for degree of normal alignment. Large abraded areas may exhibit fiber alignment unrelated to a more recent smear. Fiber melting or softening may occur as a result of rubbing or pushing against a smooth surface, such as a vehicle hood or a grazing bullet. This may also be observed as a smear; the same criteria apply to evaluating its direction of travel.

At the point of initial contact with the substrate fabric, or at the site where the object that transfers the deposit leaves the fabric, an area of light deposit is often observed, sometimes referred to as feathering. Feathering can result from a decreasing amount of material deposited toward the end of a smear, but can also be from a lighter degree of contact with the substrate. Lighter contact can occur at either the beginning or the end of a smear. Feathering is not a reliable indicator of direction unless an accumulation of material can be observed on the threads in the feathered area; otherwise, it helps delineate the line of motion but does not indicate the direction (Figure 4.14).

It can be more difficult to determine the direction of a smeared stain resulting from liquid that has been absorbed into the fabric and diffused. Liquids that are not absorbed and remain on the surface of the fabric also present challenges with regard to direction of deposit. Repelled liquids tend to form strings of beads when smeared (i.e., lines or groupings of droplets that either dry as rounded droplets or collapse upon drying but are not

Figure 4.13 Surface fibers of light blood smear aligned in direction of smear.

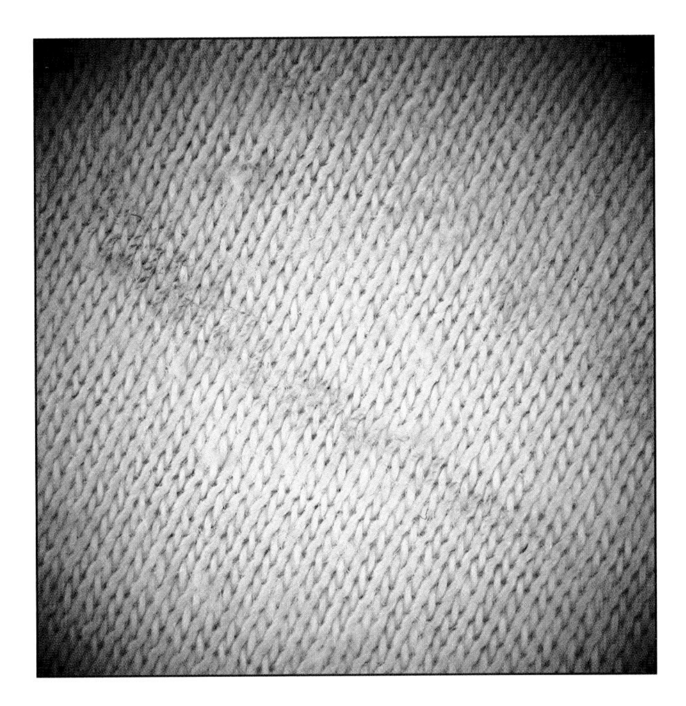

Figure 4.14 Feathering observed at the beginning of a blood smear (garment placed on curved surface, resulting in initial transfer being light contact).

absorbed). An example would be blood smeared on a plastic name tag on a uniform shirt or on a silk-screened logo on a T-shirt (Figure 4.15). It is difficult to determine the direction of these types of smears and may be impossible to distinguish them from contact deposits (Figure 4.16). In these cases, one may find that a smeared stain of uncertain direction is interrupted by creases or folds, sometimes at a seam. Clues about the direction of the smear may be gleaned by examining the direction of the creases or folds.

Figure 4.15 Blood smeared onto silk-screened T-shirt logo, partly resulting in groupings of droplets. (16× magnification)

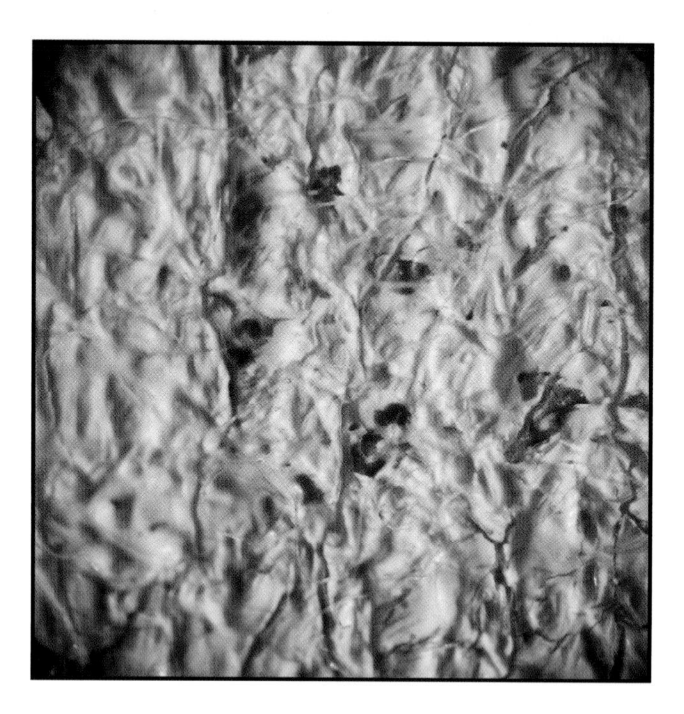

Figure 4.16 It would not be possible to distinguish these smeared deposits from drips or spatter, even at 40× magnification.

CASE EXAMPLE 4.11

Examining each deposit microscopically, examining the inside and the outside, and correlating two types of deposits

Police responded to a complaint of a burn ban violation at a beachfront property, where the residents were having a party on the beach. The beach could be reached from the house via rough unfinished wooden stairs. Words were exchanged and a scuffle ensued, resulting in superficial injuries to both the homeowner and the responding officer. The homeowner was charged with assault. His long-sleeved shirt exhibited a reddish-tan deposit that looked like diluted blood on the elbow (it was the color of blood mixed with serum). However, examination with a stereomicroscope revealed it to be an impact deposit of light-colored wood, not blood; the wood fibers were smashed. In the corresponding area on the inside of the sleeve were light reddish-brown bloodstains. The size, shape, and location of the woody deposits corresponded with a circular abraded area depicted in photographs above the right elbow of the homeowner, indicating forcible contact with a wooden surface. This lent credence to his account that the officer had pushed him. With the information about the wood deposit on the sleeve, investigators were able to examine the area near the stairs to reconstruct that part of the incident.

4.3.3 Projected Stains and Deposits — Spatters, Scatters, and Splashes

Well-developed guidelines exist for evaluating blood deposited after travel through air. Reference points for evaluation are referred to as flight characteristics, as determined from the science of fluid dynamics. Examples include the geometry and size distribution of the droplets. Flight characteristics can be used to construct hypotheses for scenarios that might result in a particular deposit. With allowances for differences in composition and density, these guidelines are generally applicable to all liquids, suspensions, and emulsions. They are discussed further in Chapter 5.

Solid materials can also be projected and are sometimes observed as scattered powders or particles. For example, if a soil-encrusted rock is thrown and strikes someone, it may leave a small, compact soil deposit and a radiating pattern of looser soil. If a mud ball is thrown, the appearance of the resulting deposit will be affected by the amount of moisture and may include some spatter and streaking.

Sometimes absence of deposits may indicate the need for further investigation. For example, in a situation where an individual was outdoors when snow and ice was on the ground, the examiner should be alert to descriptions of bruises or even broken skin that might indicate an area of initial contact. If no apparent corresponding deposits are observed on the clothing, one should consider the possibility that there was impact with a chunk of ice or with snow packed around a stone. An initial deposit of snow or ice may fall off, sublime, or melt. The latter may result in small areas of liquid suspension, which may leave particulate residue in the shape of a stain. If the initial impact is from a hard enough chunk of ice, some fabric deformation may be observed.

4.3.4 Grouped Stains, Deposits, and Damage

Groupings of the same material can form a basis for sample selection. When a pattern encompassing numerous stains or deposits is observed and samples must be selected for follow-up testing such as DNA for body fluids or FTIR (Fourier transform infrared spectrometry) for paint, it is valid to select one or two deposits from a group. The basis for the grouping should be made explicit in the laboratory notes by explanation or diagram. Groupings of stains, deposits, and damage can also assist in reconstruction of events. The groupings, organized by manner of deposit include:

1. Embedded, firmly adhering, or impressed materials, such as fibers stuck in blood or paint, can provide a time line, because the deposit of adhering or embedded materials would be contemporaneous with the wet or tacky blood or paint (Figure 4.17).
2. Accumulated deposits, such as grease and soil on the knees of a pair of pants, can provide clues to the activities of the wearer and can assist with determining whether the deposits are contemporaneous with the events under investigation.
3. Materials deposited by the same action may include droplets of blood spattered by the same blow, grease from a knife blade deposited on the margins of a stab cut, soil and mud deposited as parallel marks on one of the participants of a struggle. or tissue or fibers on the nose of a bullet from a murder scene. Deposits on the nylon jacket of a pedestrian struck by a car may include a paint smear deposited alongside a thin film of fiber melt — both would have been deposited simultaneously from sliding across the vehicle hood. Case Example 4.11 describes impacted

Figure 4.17 Dark fiber with beads of nail polish, indicating that the fiber was already on the garment when the nail polish (background) was deposited.

wood on the outside of a shirtsleeve that appeared to derive from the same event as blood deposited on the inside of the sleeve.

4. Material is often transferred from one garment to another, or from one part of a garment to another, such as mirror-image bloodstains or stains transferred through folds or layers of garments. Material transferred via a stained fabric may produce fabric impressions (Figure 4.18). This type of transfer can assist with reconstruction of events, especially if stains and deposits on the garments of one person can be

Figure 4.18 Transfer of gray clay mud from a pair of muddy blue jeans to a white shirt. Note the fabric pattern in the lower portion of the deposit.

correlated with those on another person. The basis for grouping is usually the shape and intensity of the stains or deposits. (Please see Case Examples 4.7 and 4.12.)

4.3.5 Comparing Stains and Deposits on Different Items

When comparing stains that have soaked through or transferred by contact with another stained item, it is important to evaluate how the substrate materials affect the manner of deposit, and thus the appearance, of the stains. For example, a highly absorbent cotton work shirt may be worn with a pair of polyester work pants that repels stains. Instead of reflecting the pattern on the shirt, the pants may instead exhibit beading or smears. This depends somewhat on the volume of the liquid that was transferred.

When the stains or deposits being compared are on the same item, they can simply be placed next to one another. However, stains and deposits on different items usually should not be allowed to touch. This applies especially to items from different sites, different times, or different individuals. A comparison can be performed by tracing one or both of the stains and overlaying the tracing. A photograph of a stain can also be compared this way, either by printing it onto a transparency or by using Adobe Photoshop® or a similar program that permits overlays of images. The latter technique has been reported for shoe print comparison (Griffin, 2007) and latent prints comparison and is applicable to comparison of stains on clothing.

If tracings are used, the stains can be traced with a marker onto clear plastic sheets, such as plastic photocopy transparencies. The laboratory record of the comparison should include a printed image of each stain and deposit being compared and an image of the overlay, whether using tracings, photography, or computer images (Cwiklik and McIntyre, 2002, 2003). If computer superpositions are used (reported for shoe prints; Griffin, 2007), they should be printed out. The printouts serve as a partial record of the comparison. In the case of tracings, the record should also include a photograph, scan, or photocopy of the tracing superimposed on the original, to record the fidelity of the tracing. This applies equally to computer and manual tracings.

Tracings may be superior to photographs when the range of tones in a photograph is narrow or the contrast is low. The success of the tracing method depends on tracing each discrete dot, line, or shaded area as it appears in the print. This should include each mark in the area of interest, regardless of whether it appears to be part of a pattern. Instead of "connecting the dots," one should record the "dots" themselves, because that is the actual data. If some of the data points appear to be from the background or from an overlapping stain or deposit, they can be traced in a different color. When preparing a tracing, record several reference points that can be easily recognized, such as a seam or a logo.

Tracings are also superior to overlays of scans or photographs when one stain or deposit has much less detail or much less contrast than the stain or deposit with which it is being compared, or if one reflects only a small portion of the other. In these cases, the pattern of one stain may fit completely within the pattern of another, and any differences between the two may be missed because the set of data that constitutes the pattern with less detail appears to overlap completely with the other. This is obviated by overlaying a tracing of the less detailed stain or deposit onto a photograph of the one with richer detail, or by overlaying tracings of both stains that are being compared (Cwiklik and McIntyre, 2002, 2003).

The latest computer programs can infer a shape from imaging criteria written into the program. Although these programs can be valuable, use them with great caution. In particular, it is critical to understand the criteria used by the program to interpret a signal as significant and to be aware of the signal-to-noise threshold. The bias of the image comparison program may be different than the bias of the human brain. An example from daily life is when a computer program that gives driving directions interprets a dogleg in the road as a turn. The value of tracings is to provide an objective record of what is observed. This renders the bias of the observer explicit and available for review. Unless the computer record provides a similar degree of transparency, it becomes a black box with unevaluated bias and can introduce new sources of error.

Finally, sometimes a correspondence of stains and deposits is noticed only during review of the case notes prior to writing a report. This attests to the value of a detailed diagram. If the correspondence is potentially significant and the evidence is still available, the items should be retrieved and the stains compared as discussed above.

CASE EXAMPLE 4.12

Comparing stains on different items, groupings of stains, selecting samples for DNA testing, corresponding stains noticed from diagrams in notes, and reconstruction of events

A man was bludgeoned to death in his living room. The weapons included porcelain statues as well as a variety of other objects. The identity of the assailant was not in question; he had been a guest in the victim's house and admitted to the homicide. The question was, rather, could the victim have initiated a homosexual encounter that the defendant perceived as a threat or assault? No information was derived from sexual assault kit swabs, so the clothing was examined along with scene photographs. The victim wore a knit sweater, a white shirt, and a tank top as well as a pair of black pants and undershorts. All the clothing was heavily bloodstained, with many of the stains soaking through all layers. An aggregate of stains on the upper right front of the pants, between the hip and the fly, included a stain in the right front pocket of the pants (the type of pocket formed by a flap of cloth inside), with corresponding stains on the white shirt and undershorts. These stains and their location were of particular interest because of the question of sexual activity.

Several possibilities were considered to explain the stains: (1) someone at the medical examiner's office could have transferred the blood while handling the body; (2) the defendant or victim could have reached into the pants through the pocket (essentially using the pocket as a glove); (3) the defendant or the victim could have reached beneath the pants; (4) blood from the carpet could have been transferred onto the clothing; and (5) the pants could have been open, then zipped and buttoned before the blows that produced the spatter occurred. A stain pattern analysis demonstrated that the stain on the pants pocket was deposited from the inside, on the surface that would be next to the undershorts or skin. It did not originate from inside the pocket, ruling out the bloody hand-in-the-pocket hypothesis. Nor did it originate from the outside of the pants, as the majority of the stains had, ruling out the transfer-from-the-carpet-to-pants hypothesis. In order for the stain to be where it was, either the pants had been open at some point or someone had reached into the pants with bloody hands.

The stain on the white shirt was diffuse and ran along the fibers, making it impossible to determine whether the stain originated from the inside or the outside. No blood spatter was observed around this stain, although the shirt was blood spattered above the waist level and the pants were blood spattered in the front. Thus, the shirt must have been tucked in during the blows that produced the spatter. A photograph taken of the victim at the scene, as the body was being turned over, showed the white shirt with the flap exposed and the stain visible. Thus, it was unlikely that the stain was deposited during handling of the body. The possibility that the stain transferred directly to the shirt from the carpet seemed unlikely, since no other staining was observed in that area, and was ruled out by the correspondence of the shirt stain with similar stains on the undershorts and the interior, but not the exterior, of the pants. The stain on the undershorts was also diffuse. No corresponding injury to the victim's right hip area was reported that would account for the stain being deposited from the inside. DNA testing was performed on samples of stains from the undershorts, the white shirt, and the pants pocket. All the sampled stains were attributed to the defendant (whose hands were cut from the broken porcelain), not to the victim; thus, the defendant must have reached into the victim's pants.

While writing the report and comparing the diagrams of the victim's clothing, the examiners realized that no corresponding bloodstain existed on the tank top, yet an area of apparent urine staining was recorded on diagrams of the tank top, shirt, and undershorts. Some of the bloodstains on the tank top appeared to be deposited where the shirt was folded up at the bottom and wrinkled. The yellowish stains on the shirt and tank top diagrams appeared to correspond with each other, but not when corresponding bloodstains on the upper portion of these garments were lined up. The items themselves were examined again to verify the diagrams and to reconstruct the position of the tank top. The urine stains that spread onto the shirts were probably the last deposit, occurring near death. Comparison of the relative positions of the yellowish areas and bloodstains near the shoulder and neck established that the white shirt had ridden up between deposits of the upper bloodstains and the yellowish stains. The tank top had been above the height of the bloodstain when the bloodstain was deposited on the other garments and would not have been in contact with the white shirt in the area of corresponding bloodstains that included the pants pocket.

Were the pants open during the beating? Two small droplets of blood on the inside surface of the belt were found on a bumpy leather surface that had tiny depressions in the same size range as the droplets. The two droplets on the inside of the belt were too few to form a pattern and too close in size to the belt texture to exhibit typical spatter morphology. Therefore, it was not possible to determine whether the blood was spattered or was a contact transfer, perhaps from handling the body. This area was not visible in scene photographs taken before the body was moved. Whether the pants were open at the time of the blood deposit on the inside of the belt could not be determined. In either case, someone, probably the defendant, reached under the shirt flap into the undershorts, with bloody hands or a bloodstained object, and grabbed the undershorts, producing a fold. The white shirt and pants pocket would have been included in the fold. The tank top was rolled up above this height when the stain was deposited.

Examinations of other items and of the scene photographs suggested an opportunistic homicide using decorative objects at hand as the weapons, rather than a planned murder, as initially thought. No exchange of semen or saliva was indicated. No semen was found on the victim's clothing. Had there been any semen on the defendant's long johns (thermal underwear), a thick deposit of hand cream in the fly area would have masked it; test results were inconclusive, and tests for saliva were not attempted because of the heavy cream. The location of bloodstains attributable to the defendant lent some support to the defendant's account of sexual activity despite the inconclusive results for semen on the defendant's clothing and unresolved questions about who would have originated such activity.

4.4 Sampling of Stains and Deposits

4.4.1 Basis for Sampling

Sampling decisions should flow directly from the questions in a case. The selection of samples should be governed first by significance, then by specificity and utility. Significance is defined as the potential to narrow down hypotheses. Specificity is defined as the potential to narrow sources. Utility or usefulness is defined in this context as the potential to produce interpretable results (Cwiklik and Gleim, 2009).

In Case Example 4.6, automotive paint smeared on the shoulder of a person struck by a car was instrumental in leading police to investigate the case as a homicide rather than a hit-and-run. Numerous multilayer paint chips were also observed on the clothing, only some of them with the same type of topcoat as the smeared paint. The multilayer chips had far greater potential to narrow a source vehicle for the paint, but only the less specific smears could narrow down hypotheses regarding the person being struck. The primary criterion for sampling selection should be significance. An important secondary criterion is specificity. In the case example, once the topcoat color and type was known, the multilayer paint chips with that type of topcoat could further narrow the source vehicle by means of the undercoats and primer.

If a potentially significant stain or deposit is of poor quality, it should be tested anyway, unless there is another stain or deposit that can provide the same level of information in terms of source, activity, and offense, as discussed in Section 4.2.4 (Cook et al., 1998). If the question is one of source, the more significant stain or deposit of lesser quality provides a reference for selecting additional samples that can provide better source-level information. Again, this is illustrated in Case Example 4.6, in which a smear of a paint topcoat was the reference for selecting multilayer paint chips for further testing.

A third consideration in selecting among potentially significant sample areas is the likelihood of getting useful results. An adequate sample is of sufficient size to permit the requisite testing and of sufficient quality to permit clear interpretation. Clear interpretation means that when an answer is obtained, the analyst knows which question it answers. A sample of sufficient quality truly represents the material of interest and does not include anything extraneous. Extraneous material is contamination. However, in the case of a mixed deposit or a co-deposit, a sample should represent the mixture. A sample should permit clear attribution; for example, whether the results are from the deposit or

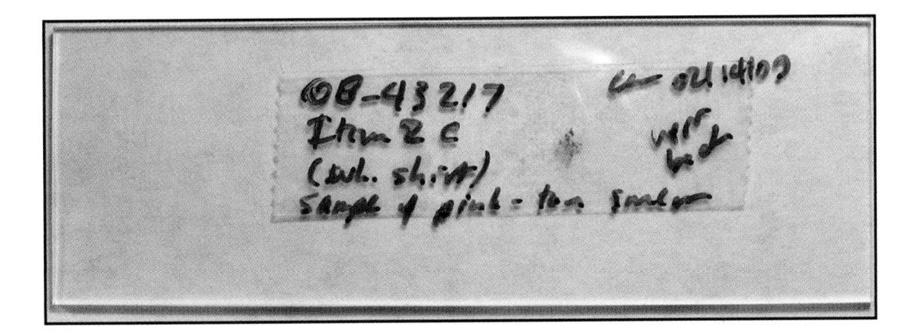

Figure 4.19 Tape lift of small portion of reddish-tan crust on a shirt.

the background, from which layer or part of a mosaic, and, when possible, from which component of a mixture (Cwiklik and Gleim, 2009).

Techniques for sampling depend on the type of material and the anticipated method of testing. For stains and deposits specifically, techniques for sampling are of several types, as described in the following sections.

4.4.2 Preliminary Sampling

Preliminary sampling using thin clear sticky-tape, (i.e., tape lifts) was mentioned earlier regarding areas where deposits would be expected but are not observed on direct examination. This method is also useful for visible deposits that are difficult to evaluate *in situ* or for deposits that would benefit from a quick examination at higher magnification. The piece of tape can be placed on a microscope slide, sticky side up, and the deposit examined using reflected light, either with a stereomicroscope or at higher magnification. The tape sample can also be mounted directly in a refractive index oil and covered with a cover slip. (Please see Chapter 8 for discussion.) To lift a visible sample from the garment, the nonsticky side can be wrapped around the end of an object, like a ballpoint pen with the tip retracted, narrow and hard but not sharp, and pressed against the deposit. This removes a small concentrated portion without tearing the tape or damaging the substrate fabric. The tape should be marked and initialed, even if it is on a microscope slide (Figure 4.19).

Preliminary samples can also be obtained using forceps, for fibers and larger particles, or for smaller or finer deposits, sharpened tungsten needle probes. A small amount of material can thus be collected and transferred directly to a microscope slide, onto a drop of distilled water or refractive index oil. This method is superior to the tape-lift method and is recommended for use by trace evidence examiners because it removes so little material that it barely alters the evidence. If the needle is tipped with a bit of adhesive, it can be swept across powdery deposits, thereby collecting a small sample that can be transferred to a microscope slide (Teetsov, 2002). (Please see Chapter 8 for a more detailed explanation.)

4.4.3 Crusts and Films

Crusts and films can be detached from the substrate using forceps and placed onto a microscope slide or piece of hard paper that is directly adjacent. Impacted fibers can behave like crusts and films and should be collected accordingly. If this technique does not work because the film firmly adheres to the substrate, it can be cut from the fabric using a razor blade. A portion of adequate size should be left in place for any reexamination. The sample

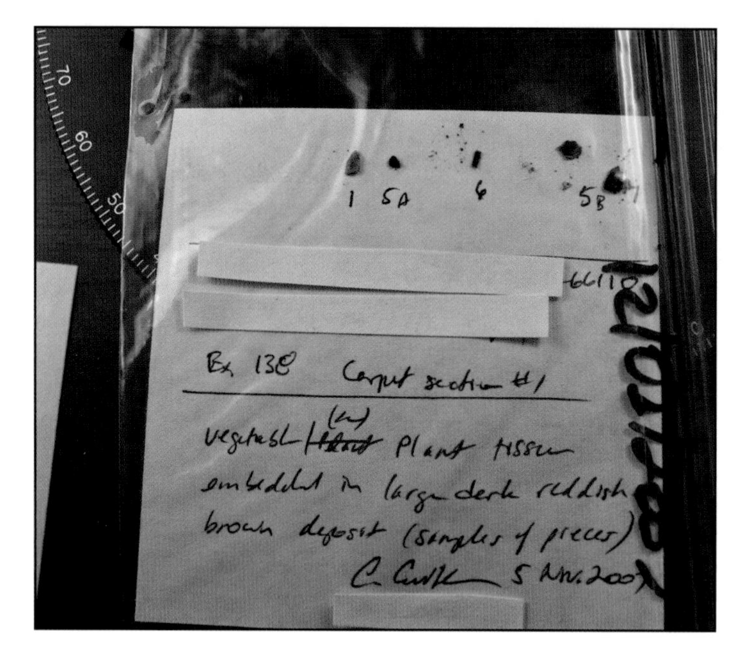

Figure 4.20 Particles and crusts removed from a thick blood deposit were placed on a sticky-note for preliminary examination, and the sticky-note was sealed in a zip-lock style plastic bag.

can be placed in a clean piece of hard paper and protected with a double three-way fold (a paper bindle). Sticky-notes such as Post-it® Notes can serve as both the sampling surface and the packaging (because they can be folded into a bindle after sampling), and they are additionally useful in providing the sample with a not-very-sticky anchor for further examination with a stereomicroscope or high-magnification reflected light microscope (Figure 4.20). Remember that the stickiness is due to an adhesive, which may be detected in FTIR or GC-MS (gas chromatography–mass spectrometry) analysis of any organics.

4.4.4 Caked Deposits and Heterogeneous Agglomerates

Caked and agglomerated deposits tend to crumble when removed and are thus best transferred using an implement with sufficient surface area to contain the material. A spatula or razor blade is usually suitable. A sticky-note or microscope slide should be placed directly adjacent to the deposit to receive the sample as well as any particles that may fly off during sampling. The samples can be placed in paper bindles and the bindles taped or placed into a secure plastic bag or paper envelope, securing any powders or small particulates from migrating out of the paper bindle. If an envelope is used, the corners should be taped.

4.4.5 Powdery Deposits

Powdery deposits, whether dense or sparse, can usually be sampled by placing a piece of paper on the adjacent fabric and tapping or gently scraping a portion of the deposit onto the paper. The paper can be folded into a bindle and taped or placed into another container. If insufficient material is recovered from a light deposit, a supplemental sample can be collected on clear sticky-tape, on a sticky-note, or on an adhesive-tipped needle probe.

The sample on thin clear tape can be mounted *in situ* and is best employed for transmitted light microscopy. Sticky-notes use a weaker adhesive that is more suitable for FTIR, GC-MS, or high-magnification reflected light microscopy. A sample collected on an adhesive-tipped needle probe can be mounted directly on a microscope slide and the adhesive dissolved. The latter technique is superior but requires more skill.

4.4.6 Stains

It is usually best to excise a portion of a stain. Disposable razor blades work well. Sometimes the examiner is not permitted to perform destructive testing, which would preclude cutting the item. In this case, one can place a petri dish (or if one is careful, a large microscope slide) beneath the stained area and drip a suitable solvent onto it, allowing the solvent absorbed from the first drop to evaporate before applying the next drop. A concentrated sample of the stain should remain on the petri dish and can be analyzed. This method should also be considered when the item must be returned intact, or to avoid causing hardship to the owner by damaging the item.

4.4.7 Viscous Deposits

Viscous deposits can often be sampled by scraping off a portion. It is best to place the sample directly onto a microscope slide. A plastic container is not a good choice because the components of the viscous material, even if water soluble, may soften the plastic or retain some of the plasticizer, compromising some types of analyses.

4.5 Questions That Can Be Addressed by Stains and Deposits

Some of the questions that can be addressed by stains and deposits overlap with those that can be addressed by traces and debris and are thus discussed in Chapter 8.

- Does this object belong here?
- How did it get here?
- Does it belong to this individual?
- How long has it been there?
- Were these items or persons in contact?
- What is the sequence of deposits?
- What path do the deposits indicate?
- In what order did things happen?
- Does it fit with the story?

Another question, more often encountered in examining deposits, is whether contact was forcible. In other words, was a grease deposit ground into or simply touched to an item of clothing? Was a paint smear deposited by impact? Was someone struck with a rock, leaving deposits and damage? Was a piece of fruit smashed against someone, or were the innards projected after hitting something else? Was a plastic auto part cracked by someone being thrown against it, or by striking it then sliding across? The clues may be in deposits and damage on the clothing as well as on the damaged part. For example, plastic smears (or paint smears, if the plastic is painted) would be expected in the latter scenario, but only minimally in the former.

4.6 Sorting Tools for Stains and Deposits

4.6.1 Sorting Tools for Preliminary Evaluation

What does it look like (type of deposit)? What is it made of (type of deposit)? How did it get there (manner of deposit)? What happened to it (alteration and wear)? If it is not what it first appears to be, what are some other possibilities?

Consider a thin white translucent crust that looks like a crust of dried semen, yields a negative result with acid phosphatase (a preliminary test for semen), and exhibits no spermatozoa. Could this be an older, degraded stain from an aspermic donor? Should it be tested anyway for DNA? Other possibilities include paste, wood glue, crusts from cooked starch, and other body fluids such as vaginal secretion. If one of these can be corroborated or supported, there is no reason to think it might be a degraded deposit of semen. If plant cell structures are observed under the stereomicroscope, it might be oatmeal or other starchy food residue (Figure 4.21). If numerous epithelial cells are observed (this would require higher magnification), it could be from a mucous membrane such as the vaginal vault or nasal passages. (Suggested terminology for describing stains and deposits is presented in Section 4.10.)

Remember the game where one person thinks of something that could be pictured and others try to guess what it is by answering questions? The first question is always, "animal, vegetable, or mineral?" We would now have to ask, "animal, vegetable, mineral, or synthetic?" That question is a useful sorting tool for narrowing down a type of material observed under the stereomicroscope. It is a useful question even when it cannot be readily answered, because it forces the examiner to think about the defining characteristics of the components of a stain or deposit.

Similarly, we want to know how the deposit got there, whether by contact or projection, by direct or indirect transfer, and whether it exhibits motion and direction and is related to other deposits and damage. We want to know about any alteration or wear and whether damage is recent or not.

Sorting tools are further described in Appendix 4. The sorting tools and terminology are not a rigid classification system. Each examiner should use those that are most useful.

4.6.2 Sorting Tools for Examining Samples Received from Another Examiner

When a sample is received for further testing from the scientist who has examined the clothing, the sample should be described by the person receiving it. Not only does this provide a record of what was received, but it also provides enough information so that subsampling can be documented and results of testing attributed to a known part of the deposit and to a component or location that has been described.

4.7 Establishing a Reference Collection

The best way to learn how to recognize common materials and how they are deposited is to examine exemplars. These can be prepared easily and kept in the laboratory as a reference collection. In addition, the examiner will encounter materials in daily life, including materials in context, and will see materials being deposited or altered. These can be brought to the laboratory and studied as well. (Common types of materials that can form the basis for a reference collection and suggestions for preparing exemplars are included in Appendix 3.)

Figure 4.21 (Top) White crust on blue jeans, produced by oatmeal. (Bottom) Apparent plant structure in the crust. (25× magnification)

4.8 Writing Reports

Most agencies have report formats that examiners are required to follow. When reporting examinations of stains and deposits, give special consideration to the following issues:

- Provide a description of the garment that gives the reader a mental picture. Police detectives, investigators, or attorneys who view the items prior to trial may make their own observations and form their own questions about the evidence. Descriptions in the scientific report should be written with this in mind, especially if the stains and deposits are not as they appear to the naked eye.
- Document the relationships of stains, deposits, or damage that appear on the same item or on items of clothing worn at the same time.
- Note which stains, deposits, and damage may be unrelated to the incident under investigation.
- Report on the direction of deposits and damage if it provides useful information.
- For clarity, indicate the reported source of the item (e.g., "blue long-sleeved dress shirt reported to be from John Doe"). If the source is not documented in the identifying labels or paperwork, the examiner should seek out the information.

CASE EXAMPLE 4.13

Reporting of stains and deposits used in reconstruction of a vehicle accident

A vehicle struck a girl who had darted into the road, flipping her over the hood and onto the pavement; the driver had tried to avoid hitting her, as evidenced by his striking a mailbox and leaving tire tracks on the berm. A neighbor ran to assist the girl and was struck by a second vehicle. The girl died and the neighbor was injured. Whether the girl had also been run over by the second vehicle was the subject of laboratory work. The driver of the second vehicle was charged with negligent homicide. No damage attributable to the undercarriage of a vehicle was observed on the girl's clothing. Excerpts from the laboratory report follow:

A: Unrelated bloodstains

The upper clothing of [the girl] is extensively bloodstained on the back (where it pooled) and on the right side, with lesser staining on the left side and front. The T-shirt is extensively blood-soaked; however, this garment is depicted in autopsy photographs with both front flaps twisted at the sides of the deceased and lying on the autopsy table, where pooled blood would soak in. The sweatshirt is also depicted partly folded under the right arm.

B: Related and unrelated deposits

Extensive deposits of soil observed on the lower legs of the blue jeans are accompanied by small deposits of oil or grease; abrasions were observed near the hems. White and bluish-gray paint observed on the blue jeans and shoe were deposited while wet and are unrelated to this incident.

C: Unrelated deposits; characterize activities of the wearer

The clothing of [the neighbor] is bloodstained mostly on the left side with blood dripping across the chest and on the right sleeve. It is grease-stained in areas that would typically be grease-stained from working on cars. His work pants are heavily grease-stained especially in the back on the thighs and buttocks, but also in the upper front. His undershorts are grease-stained in the margins of the fly and there is grease in his pants pockets. These grease stains appear to be from normal occupational use unrelated to the incident under investigation. The lower pants legs are splashed with mud. Damage related to impact is described below.

D: Deposits are of a different color than the component paint chips

The gray deposits on the left shoe include tiny chips of black or dark gray paint and gray smears that appear to result from a melding of the white plastic of the shoe with black or dark gray paint.

E: Deposits are of a different color than component particles

The green marks associated with abrasions to both hips of the blue jeans (near the right and left pockets) are not paint; the green color is produced by a combination of slightly yellowish particulate material associated with the fabric abrasions and the blue fibers of the blue jeans. This is visible as green to the naked eye and at lower magnifications.

F: Marks on garment are scrapings, not deposits; related damage

Vertical blue striations on the rear pockets of the pants are not deposits, but are heavy abrasions which by scraping off the heavy grease deposits have exposed the blue fabric below. There is associated tearing of the bottom seam of the right rear pocket.

4.9 Summary

An objective analysis of stains and deposits, even when preliminary, can assist in selecting samples for further examination, in answering questions in a case, and in reconstructing events. Even when a preliminary analysis does not stand alone, it can provide clues for investigators to follow. A rigorous preliminary analysis requires little equipment other than a good stereomicroscope. A range of analytical instrumentation can be used to conduct further testing of the materials involved. One need not be a specialist in trace evidence or chemistry to notice soil or paint, or in blood pattern analysis to notice blood spatter, or in firearms to notice gunpowder residue. A specialist in another forensic discipline can conduct a general preliminary evaluation. Many a case investigation has been furthered by an alert examiner who paid attention to clues, even if they were of an evidence type not within the examiner's specialty. Once a scientist has systematically studied stains, deposits, and damage via reference collections and experiment, he or she can provide more comprehensive information in casework.

A final case example follows:

CASE EXAMPLE 4.14

Reconstruction of events using stains and deposits

The body of a man was found in a wooded area not far from a roadway. His pants were fastened but pulled down below his knees. A man and his uncle confessed to abducting the victim at the behest of the man's lover, who wanted the victim roughed up. They picked up the victim in a rented car from a location in the city, ostensibly to give him a ride home, but instead took him to the wooded area. The uncle claimed that he was tending to the car when his nephew killed the victim instead of just roughing him up. Was the victim killed at the scene, or was he taken there after being beaten or killed at another location — perhaps a site associated with the victim's lover? How was the uncle involved?

Blood spatter was observed on the victim's pants, but not his shirt, indicating that another outer garment had been worn during the beating (the nephew said that he burnt the victim's jacket). Mud smears and abrasions on the pants and shirt indicated that the victim was dragged, head first, while his pants were already down. Autopsy photos depicted abrasions on the buttocks produced in the same direction as the mud smears, but no abrasions were observed on the adjacent waistband. If the pants had slid down while the victim was being dragged, abrasions would be expected on and near the waistband. Who pulled the pants down? The nephew's finger was cut to the bone in the initial struggle when the victim pulled a knife. Had the nephew then pulled the pants down, his blood would be expected on the waist, belt, fly, or hips, but there was little blood of any type in these areas. Unless the pants were down before the knife was drawn, which is unlikely unless the victim already had the knife in his hand, someone other than the nephew pulled down the pants.

The soles of the victim's boots were clean except for soil in the crevices and bits of grass on the side soles, suggesting that he walked on wet grass but not in the mud where the body was found — it was not raining that day, but the ground was wet. This suggests that the victim was able to walk when he arrived at the grassy area nearer the road, so his pants would have still been on. This was corroborated by soil deposits on the front of the shirt and pants that indicate the shirt was tucked in when the soil was deposited. Someone probably pulled his pants down to inhibit his movements during a struggle. Once he was unable to resist, the victim was dragged deeper into the woods. Scene photographs showed a grassy path to the spot where the body was found. Several flat pieces of broken asphalt were found during a site visit. This would have accounted for the scratches to the buttocks and to black tarry deposits on the front pants pocket, where it would have been flipped up and exposed (he was dragged with his weight on one leg, thus partly on his side). At this stage, he would have been unable to wield a knife to cut the nephew. It is likely that the uncle assisted the nephew, because someone other than the nephew most likely pulled the pants down. This would have occurred closer to where the car was parked, not deeper into the woods.

Death was by strangulation. The turtleneck shirt was bunched up in the back, the fabric was distended, and blood attributable to both the nephew and the victim was deposited in the bunched area. The belt to the pants was found next to the body. Was he strangled by someone pulling on the shirt? Or could the victim have been dragged by the belt around his neck, then strangled? No traces of belt leather were found around the turtleneck collar, but the possibility could not be eliminated. There was no evidence to suggest that another site was involved. A deeper involvement of the uncle in the beating was strongly supported, but the strangulation may have been incidental to the beating. This could not be established.

4.10 Terminology for Stains and Deposits*

4.10.1 Terminology for Appearance

If deposited as a liquid, the resulting stain may be:

Dried stain: Soaked into fabric, usually with well-defined edges
Particulate residue: Particles deposited in a defined ring or concentric rings
Oily stain: Diffuse edges, often with translucent appearance; includes components of materials such as wood stain or some facial cosmetics; may result from food oils, automotive oil, hair oil products, or medicinal oils
Bleached area: Fabric dye loses color partly or completely; visible in individual fibers; partially bleached areas on a black fabric may be pinkish

If deposited as a solid, the resulting deposit may be:

Powdery: Fine, loose particles of fairly uniform size
Sandy: Coarse, loose particles of fairly uniform size
Crystalline: Composed mostly of crystals
Aggregate: Particles of different types or sizes bound together, as in concrete
Clumped: Particles held together by moisture or other loose bonds, as in damp soil
Caked: Thick layer of loosely bound particles
Fibrous: Length much greater than width
Particulate: Length versus width differences not pronounced
Firmly adhering debris: Particles and fibers in close contact with the substrate
Loose debris: Particles and fibers in light contact with the substrate

If deposited as something in-between a solid and a liquid (dried or polymerized, suspensions or emulsions), the resulting deposit may be:

Film: A flexible continuous layer, such as paint or nail polish
Crust: A brittle continuous layer, such as glue or visible semen
Glaze: A shiny transparent crust
Chips: Thin, small, fractured pieces of a film, hard plastic, metal, or other hard object
Flakes or crusty scales: Thin, small pieces of brittle films or broken crust
Gelatinous: Somewhat soft, often glistening
Glutinous: Looks like fat, glistening and amorphous
Globules: Soft spheroids, such as clumps of starch grains
Greasy: Viscous, low-vapor-pressure material, often with suspended solids; spreads easily
Tacky: Not completely dry; may be sticky
Rubbery: Pliable and tough film or particle, readily stretched and deformed but difficult to break apart; abrasion results in rubbery curls of a wide range of sizes that vary with the source material

* This material is part of a course in clothing examination taught by the Pacific Coast Forensic Science Institute (used with permission).

4.10.2 Terminology for Manner of Deposit

Contact deposits

Direct contact: Deposit directly from the source of the material

Contact transfer: Deposit on one object or person transferred to another object or person or to another part of the same object or person; a form of indirect transfer; examples include blood transferred through a fold or soot transferred from someone's hand to a garment

Indirect transfer: See *Contact transfer*

Smears and drips

Smears and drips: Stains or deposits produced or altered by motion across a surface

Drip: Liquid, emulsion, or suspension that ran or dripped down a surface after the original deposit, via projection or contact

The IABPA (International Association of Blood Pattern Analysts) and SWG Stain guidelines (see Chapter 5) suggest the following to describe smears of blood; paraphrased here to be generally applicable:

Swipe pattern: Contact transfer of material during motion

Wipe pattern: A pattern produced by motion through an existing stain or deposit, removing material and/or altering its appearance

Projected or propelled

Impact deposit: liquid or suspension: Dispersion of liquid by applied force, such as by the tire of a vehicle that splashes mud onto the legs of bystanders; includes splashes and spatter

Spatter: Liquid, colloid, or emulsion propelled through the air as drops

Splash: Liquid, colloid, or emulsion propelled as a larger volume

Aerosol: Fine mist of extremely fine droplets light enough to be partly suspended in air

Passive drop: Fell with the force of gravity alone

Impact deposit: solid: Dispersion of a solid projected onto another surface, such as a clump of soil or snow-covered ice striking a person and scattering material in smaller pieces

Scatter: Dispersion of a solid by impact, resulting in radiating distribution of particulate material; see *Shatter*; shattering glass and crystals do not exhibit radiating distribution

Shatter: Shards of glass or pieces of crystal propelled by sudden rupture of chemical bonds

4.11 References

Cook, R., Evett, I.W., Jackson, G., and Jones, P.I., A hierarchy of propositions: Deciding which level to address in casework, *Sci. Justice*, 38(4), 231–239, 1998.

Cwiklik, C., An evaluation of the significance of transfers of debris: criteria for association and exclusion, *J. Forensic Sci.*, 44(6), 1136–1150, 1999.

Cwiklik, C., *Describing Stains and Deposits: Suggestions for a Standard Terminology*, Presented at Inter-Micro 2006, Chicago, July 2006.

Cwiklik, C.L. and Gleim, K.G., Forensic casework from start to finish, in *Forensic Science Handbook*, Vol. III, pp. 1–30, Saferstein, R., Ed., Prentice-Hall, Englewood Cliffs, NJ, 2009.
Cwiklik, C. and McIntyre, L., *A Simple Tracing Method for Comparing Prints and Stains*, Presented to the Northwest Association of Forensic Scientists (NWAFS), Coeur D'Alene, ID, October 2002.
Cwiklik, C. and McIntyre L., *A Simple Method for Comparing Difficult Prints and Stains*, Presented to the American Academy of Forensic Sciences (AAFS), Chicago, February 2003.
Griffin, H.R., *Adobe Photoshop Court Demonstrations of Footwear Impression Evidence*, The International Association for Identification's 92nd International Educational Conference, San Diego, CA, July 22–28, 2007.
McGarvey, L. and Cwiklik, C., *Interpreting a Group of Bloodstains Assisted by a Simple Overlay Tracing Method*, Presented to the Northwest Association of Forensic Scientists, Coeur D'Alene, ID, October 2002.
Petraco, N. and Kubic, T., *Color Atlas and Manual for Microscopy for Criminalists, Chemists and Conservators*, CRC Press, Boca Raton, FL, 2003.
Plant disease patterns: http://plantpath.ifas.ufl.edu/pdc/Inclusionpage/aboutv.html; http://www.hortnet.co.nz/publications/hortfacts/hf205021/disgloss.htm; accessed 2006.
Soil Survey Staff, *Keys to Soil Taxonomy*, 10th ed., USDA, Natural Resources Conservation Service, Washington, DC, 2006.
Teetsov A.S., An organized approach to isolating and mounting small particles for polarized light microscopy, *Microscope*, 50(4), 159–168, 2002.

Pattern Evidence 5

Patterns found on clothing may be used with either reconstruction or individualization techniques. Reconstruction patterns can help reconstruct past events; blood patterns, gunshot residue, and clothing damage are examples. Blood patterns and gunshot residue are discussed in this chapter and clothing damage is discussed in Chapter 6. Individualization patterns are those that can be associated with the item or person responsible for it, such as fingerprints, footwear, and tire marks; these are discussed later in this chapter.

5.1 Blood Pattern Analysis (BPA)

Patterns of bloodstains on clothing are some of the most important patterns of evidential value in a violent crime; they are also probably the most common type of reconstruction pattern. Blood pattern analysis is a recognized field of forensic expertise and has found particular application at crime scenes. The interpretation of blood patterns on clothing has great limitations due to the nature of clothing (more absorbent surfaces than walls or floors, for example) and the fact that the clothing of a suspect or victim may move after or during the shedding of the blood. Nevertheless, the finding of blood patterns on clothing should lead the examiner to attempt an analysis of the patterns and their possible source(s), if appropriate. This may lead to hypotheses, to submitting the clothing for further examination by a specialist, or to no conclusion due to the limitations of the examiner, the evidence, or both.

A preliminary interpretation of bloodstain patterns may also serve as a basis for selection of samples for DNA testing. If more than one person was injured, a reconstruction may be predicated on knowing whose blood was spattered where.

If spattered blood is found on clothing, it may place the suspect at the scene of a violent altercation. The mere fact of blood on the clothing may not be the most important evidence in a violent crime case. The pattern or appearance of that blood may impart useful information as to the method of deposition. Interpretation of the blood patterns must also be done in conjunction with the knowledge of whose blood was shed. Multiple blood sources may significantly affect interpretation (see Chapter 7).

Bloodstain patterns on the clothing of a victim are usually that of the victim, but occasionally there will be blood from the suspect in cases where a violent altercation has occurred and the suspect has been injured. Any difficult-to-explain features of bloodstain patterning, or inconsistent directionality, may represent the blood of the assailant.

Many case studies illustrate that the interpretation of bloodstain patterns on clothing often involves supporting or refuting the suspect's version of how his or her clothing became stained with blood. Conversely, the interpretation may need to explain why there was no bloodstaining on a suspect's clothing despite stated contact with the victim. The Sam Sheppard case in the United States is a famous example of conflicting theories of blood deposition on clothing in a criminal case, and the controversy lingers some 60 years after the murder of his wife (Inman and Rudin, 2001). The McLeod-Lindsay case described below is a notable example from Australia. Suspects may claim that they came into contact with the victim only after the injuries were inflicted. The source of the blood on the suspect's clothing was not an issue in either of these cases; it was the method of deposition that was in contention. Blood pattern analysis may thus be vital in cases where the victim knows the suspect.

The question that blood pattern analysis attempts to answer is, How was the blood deposited on the clothing? This deposition generally falls into one or both of the following categories: *passive* bloodstaining, including transfer, flow patterns, saturation stains, and stains resulting from dripping blood; or *active* bloodstaining, including impact spatter, arterial spurts, expirated bloodstains, and cast-off (James et al., 2005).

The principles of active bloodstain pattern analysis rely on the physical laws followed by droplets of blood falling or projected through space. Spatter is created when sufficient force overcomes the surface tension of the blood. Blood spatter may be produced on clothing by an impact mechanism or a projection mechanism. Impact spatter is the direct impact of an object with blood or the impact of blood with a surface that produces satellite or secondary spatters. Until recently, impact spatter patterns were classified according to the correlation between the velocity of the object striking the blood and the size of the resulting spatters; that is, low-, medium-, and high-velocity impact.

Low-velocity impact spatter patterns are formed where gravity is the only force acting on the blood. These are typically dripping patterns, such as blood dripping from the nose onto the clothing below (passive bloodstaining). The primary stains are generally 4 mm or greater in diameter.

Medium-velocity impact spatter results when moderate force from some object causes pooled blood to scatter in all directions surrounding the contact. Stomping a foot or shoe into pooled blood will cause medium-impact blood spatter on the lower legs of the trousers. The diameters of the resulting spatters range from 1 to 3 mm in size, although smaller and larger stains may be present.

High-velocity impact blood spatter results from high force acting on a blood source. Gunshot and explosions are examples. Impact spatter associated with gunshot may produce minute spatters of blood that are less than 0.1 mm in diameter and appear like mist. However, they may exhibit a wide size range, up to several millimeters or more.

Recently, it has been recognized that there may be overlap among all three impact spatter patterns (James et al., 2005). A single mechanism may create spatters of sizes that fit all categories. The mechanism that creates high-velocity impact blood spatter may create bloodstains of a variety of sizes that would fit into the categories of impact spatter associated with beating and stabbing events as well as spatter produced by expiration of blood. Thus, this book uses the terminology of impact spatter and projected spatter in general, without denoting the velocity.

Bloodstains are classified by their geometric appearance and taxonomy. The three main bloodstain categories (James et al., 2005) are *passive* (drops, transfer, and flow patterns), *spatter* (impact, secondary, and projection mechanisms), and *altered* (clotted, diluted, diffused, insects, and voids).

The structure and composition of the fabric of the garment will influence the appearance of any bloodstain on the garment. Clothing does not have the smooth, nonabsorbent surface of walls and floors, and so a bloodstain may be distorted in shape from the spheres and ellipses observed at crime scenes. This distortion will depend on the ability of the fabric to absorb the bloodstain and the looseness of the weave, knit, or felt. For example, a passive blood drop from a nose bleed may appear almost spherical on a weave such as

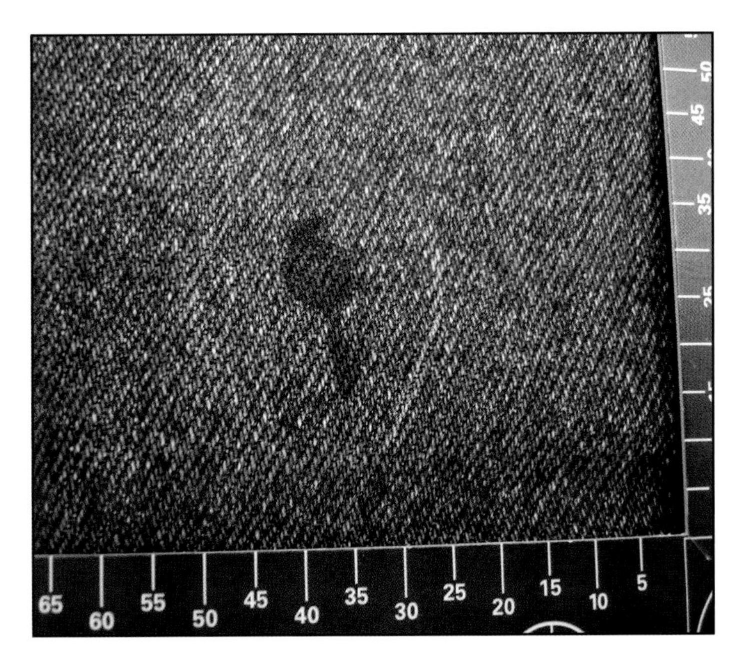

Figure 5.1 Blood drop/drip on a pair of denim jeans.

those composing business shirts, but may appear distorted on a very loose knit. Figure 5.1 shows a passive blood drop/drip on a pair of denim jeans; the direction of travel is toward the bottom of the page. However, the material of the garment precludes any definitive directionality of the blood through observation only. Figure 5.2 depicts a passive blood drop from a bloody nose on a plain knit T-shirt. The material of the T-shirt is thinner than that of denim jeans, and the heavy blood drop on the T-shirt has a denser center but a round periphery. Consequently, passive blood drops may appear different in morphology on different clothing fabrics, even though the fabric is tightly knitted or woven, due to the varying thickness of the material.

The direction of motion of a bloodstain on clothing can usually be determined from the shape of the stain; the tapered, elongated end of the stain points in the direction of travel. Identifying which surface of a garment (inner or outer) the bloodstain initially impacted is straightforward if the blood has not soaked through the fabric. However, when penetration of the blood occurs, the origin surface may not be clearly apparent, and careful observation, preferably under magnification, will be necessary. It may be advisable to perform tests on the material to support the conclusion.

Some fabrics, especially those treated to be water repellent, show little or no blood absorption, and dried blood droplets may even fall from the material during transportation (Slemko, 2003). Due to the numerous variables involved in interpreting the angles of impact of bloodstains observed on fabric, especially when some factors are unknown, we do *not* recommend attempting to interpret impact angle.

It is important to ascertain whether the bloodstains on the clothing were produced through the crime event or by subsequent actions of police, medical personnel, or a pathologist. Stains may be produced during handling of the body at the scene, transportation, or autopsy. Crime scene notes and photographs, including those of the autopsy, should be examined and compared with the current clothing condition.

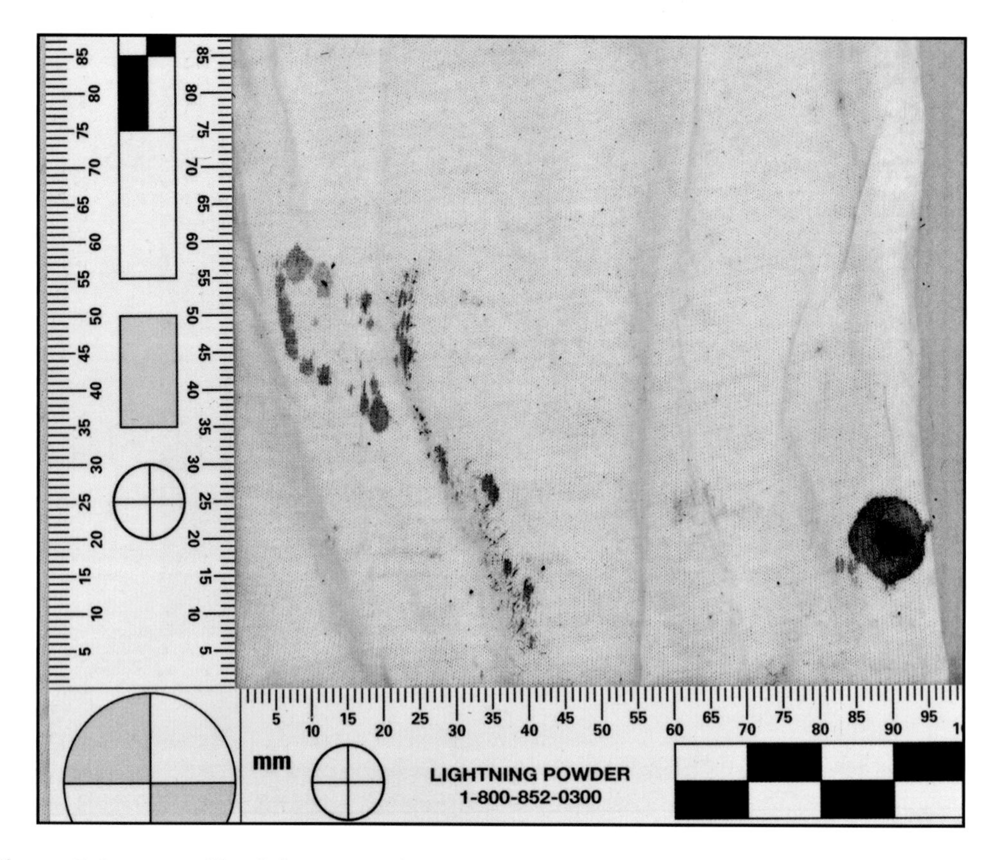

Figure 5.2 Passive blood drop on a white, plain knit T-shirt.

As far as practicable, ensure that any blood spatter interpretation is performed before sampling the blood to determine origin. In any case, never interpret samples once they have been removed from the garment. The following case study (Chisum and Turvey, 2006; Teichroeb, 2004) illustrates the consequences of (now) outdated sampling procedures and interpretation:

> Two men were arrested for the execution-style slaying of a convenience store clerk in the early 1980s. The prosecution alleged that there was high-velocity blood spatter on both men's clothing, showing that they were in close proximity to the victim when he was shot. It was also alleged that two flakes of gunpowder were found on the clothing of one of the accused, and it was this evidence that persuaded the juries. The two men were convicted in 1986 and 1987 and each sentenced to life imprisonment. However, DNA testing in 1994 determined that all but one of the 10–12 remaining blood spatter particles from their clothing did not match the victim. It was discovered that the clothing examiner had used the same ruler to scrape both the victim's and the suspects' clothing on the same day; the examiner also did not wear gloves. Not recognizing the possibility that the clothing may have contained blood spatter, the clothing examiner scraped and vacuumed the surfaces and then microscopically examined the debris. The examiner concluded that the debris contained particles of a shape and size consistent with high-velocity blood spatter. A tip-off in 1994 led police to another man, who matched a fingerprint found on tape at the crime scene. This man confessed before subsequently killing himself. The initial two accused were released from jail.

The sampling procedures in this case may have led to cross-contamination between the clothing items. The one blood particle from the suspects' clothing that matched the victim may have originated through contamination from the victim's clothing. Moreover, it is not possible to denote the original size or shape of a bloodstain deposited on clothing once that clothing has been scraped or vacuumed.

5.1.1 Impact Blood Spatter

Impact spatter results from an object directly striking a source of exposed blood. Resultant stains may range in size from 0.01 to 3–4 mm or larger in diameter. There should be further case-dependent analysis to associate the spatter pattern with beating, stabbing, gunshot, or explosive events.

5.1.1.1 Gunshot

Gunshot cases involve two sources of impact spatter: *backspatter* of blood is associated with an entrance wound to the body, whereas *forward spatter* is associated with an exit wound. Backspatter may be found on the weapon and the shooter, especially in the hand and arm areas if the shooter is close to the victim. These areas of the deceased should also be inspected if suicide is hypothesized. The cuffs and sleeves of shirts and hand jewelry such as rings and watches should also be examined for bloodstaining if backspatter is suspected. Although backspatter may be found as far up the shooter's sleeve as the shoulder, it is usually limited to the cuff or lower sleeve area (Pex and Vaughan, 1987). In addition to blood, backspatter of brain tissue, fat, muscle, bone fragments, skin, hairs, and even ocular tissue has been documented in casework (Karger et al., 2002). The maximum distance of blood backspatter has been estimated in the literature as 4–5 feet (James, 2007). The small droplets of blood typically seen in a "mist" produced by gunshot generally do not travel far due to air resistance (similar to the principle that large drops of blood travel further than small drops, when subjected to the same mechanism). As discussed previously, microscopic observation and appropriate illumination of the bloodstains on the garments may be required, especially if they are produced through gunshot, due to the small size of some of the stains.

5.1.1.2 Beating and Stabbing

Beating and stabbing with implements are other means for causing impact blood spatter. The first blow rarely produces spatter; it is usually the exposed blood receiving the impact that produces the subsequent spatter. Although assailants often receive significant blood spatter on their shoes or clothing, sometimes little or none is present. The quantity of spatter will depend on the relative position of the participants and the number and location of blows. Even if the assailant washes his or her clothing to hide involvement in the offense, other items that were worn that may contain spatter may be overlooked. Socks, belts, hats, glasses, watches, rings, earrings, and other jewelry should be inspected for spatter.

Figure 5.3 depicts the left front shoulder area of a white woven shirt. An impact spatter pattern consisting of blood drops less than 1 mm in diameter is present.

All areas of clothing should be inspected carefully for impact spatter. Areas that are not obvious targets for spatter may nevertheless have exposed surfaces that might have received impact blood spatter, such as the inside of a cuff or the inside of a pocket.

Figure 5.3 Left front shoulder of a white woven shirt with a blood impact spatter pattern of drops less than 1 mm in diameter.

5.1.2 Projected Blood Spatter

Projected spatter is created as the result of a force other than impact. It may result from expirated blood, cast-off blood from swinging of an implement coated in blood, or an arterial spurt or gush.

5.1.2.1 Expirated Blood

Expirated blood from the mouth or airways of the victim may produce a distinctive projected pattern. Air bubbles may appear in the stain patterns on the clothing of the victim or indeed another individual if they are in close proximity. Expirated blood showing air bubbles among a large amount of contact bloodstaining is pictured on a plain weave pillowcase in Figure 5.4. The victim in this case was bludgeoned to death in his bed.

It should be noted, however that air bubbles are not always present in an expirated bloodstain pattern. The victim must have blood in the mouth or nose or some type of injury to the chest or neck that involves the airways. The bloodstains may also appear diluted if mixed with sufficient saliva or nasal secretions. The following case study illustrates the problems encountered when expirational spatter was not recognized or considered as a cause of the blood patterns (Brown and Wilson, 1992):

> Alexander McLeod-Lindsay arrived home in the early hours of one morning in 1964 after working at a New South Wales hotel as a waiter. He explained that he found his four-year-old son severely injured with a fractured skull in the hallway, and there was no lighting. He grabbed a torch and discovered his wife just inside the door of the bedroom, naked except for her pajama top and covered in blood. He tried to pick her up, but she had extensive injuries and he rushed to the neighbors for help. His wife survived, but she was brain damaged and lost the sight of her right eye. She had been bashed with a pick from a jackhammer that belonged to Alexander and was found in the foyer of the house. Bloodstaining on his gray-blue wind jacket resembled the impact blood spatters on the wall and a wardrobe, and the prosecution suggested that he had gone home earlier from his job, since he lived only a few

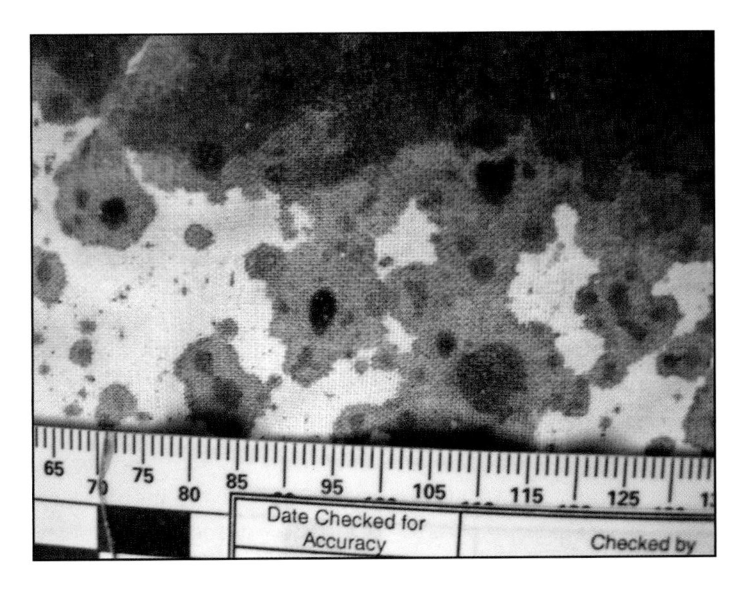

Figure 5.4 Expirated blood showing air bubbles among contact bloodstaining on a plain weave fabric.

minutes from the hotel. Bloodstains were found on the front lower part of his waiter's black trousers and extended to a point above the knees; they were also present on his black shoes and black bow tie. It was alleged that he bashed his wife during the short time interval of leaving his work and returning; his wife did not believe that he was her assailant. He was found guilty and sentenced to 18 years in jail, and in 1969 an inquiry upheld his conviction. He was released on parole in 1973, and in 1990 the New South Wales Supreme Court directed there be another inquiry. A blood dynamics expert visiting from the United States in 1988, Anita Wonder, had noticed from the photographs that there were blood clots on the sleeves of the jacket. Further experiments by Dr. Tony Raymond, a forensic scientist from Australia, showed that coughing or sneezing blood could be a source of many of the stains on McLeod-Lindsay's jacket (Tony Raymond, personal communication). There was an absence of bloodstaining in the midriff area of the accused, which did not accord with the prosecution's impact spatter theory. Justice Loveday found that the staining on the jacket could be explained by a combination of arterial damage and expiration. Blood on the trousers, where spots of fairly uniform size had run up the leg to the knee, was more problematic. One explanation was that these spatters had originated at the hospital, where he tended to his wife on the hospital trolley. Justice Loveday considered there was reasonable doubt as to guilt, and Alexander McLeod-Lindsay was subsequently pardoned.

A paper written in 1967 by one of the original examiners shows the advances today in blood pattern analysis (Merchant, 1967). It is now recognized that different mechanisms of impact spatter may produce similar patterns of blood spatter and that limitations apply.

5.1.2.2 Arterial Spurt

Another example of projected spatter is that caused by an arterial spurt onto the clothing of either the victim or the assailant. An arterial spurt pattern results when an artery is cut or severed; repeated spurts cause a characteristic pattern corresponding to the diastolic and systolic pressure in the heart. Arterial spurts will arise only if the artery is not covered by clothing. The arms, legs, and neck have arteries close to the surface that may spurt blood when cut. This spurt may resemble an electrocardiogram tracing. The individual drops tend to be large.

5.1.2.3 Cast-Off Spatter

Cast-off patterns result when a bloody object, such as a hammer, is swung through space and throws off droplets onto the clothing. Blood adheres in varying quantities to the object that has caused the injuries. A centrifugal force is generated as the bloody object is swung through the air in repeated blows at the victim. Blood is flung from the object and forms a cast-off bloodstain pattern. These patterns appear linear, but their size, distribution, and quantity may vary. An examination of the crime scene photographs may indicate whether cast-off patterns were present on the walls or ceiling, the most common sites for these patterns. Patterns may also be detected on clothing, especially on the rear of the shirt or back of the trousers of the offender. Figure 5.5 shows the back of a T-shirt from an offender who stabbed the victim more than 30 times, causing severance of the carotid artery and jugular vein. A cast-off pattern can be observed on the left upper back, with spatter to the left and below.

When cast-off patterns are on the front of the clothing, this may indicate mere proximity rather than actual participation in a beating or stabbing event.

5.1.2.4 Secondary Spatter

Secondary spatter or satellite spatter occurs when blood drops fall into a preexisting pool of blood, resulting in a distribution of spatter around the periphery of the parent stain. The size range of satellite spatter overlaps the size range of the spatter produced by other spatter mechanisms and should always be considered when evaluating clothing. Suspects often claim to have shaken or attempted to resuscitate the victim upon "discovery" of the body, causing spatters on their clothing.

Satellite spatters have a horizontal and a vertical perspective. As each drop hits the liquid surface, it can cause small droplets to splash upward, and some of these may hit a nearby vertical surface. Cuffs of trousers that have been located for a time near a blood pool onto which drops were falling are an example. The secondary droplets splash upward, so

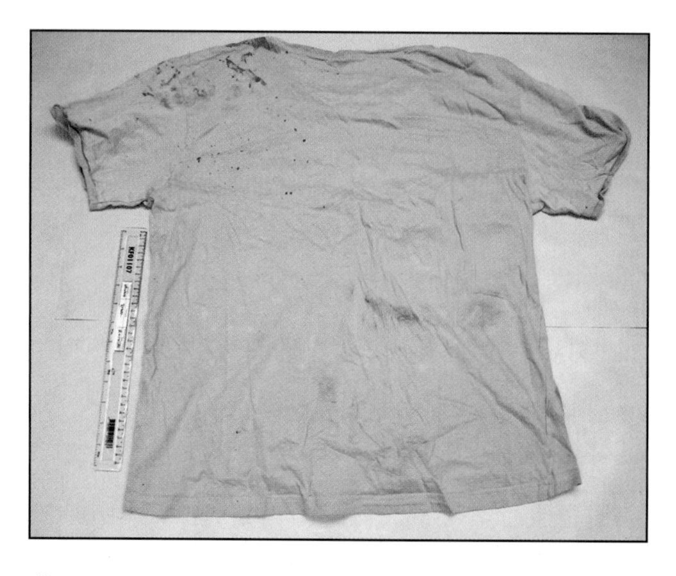

Figure 5.5 Cast-off pattern on the upper back of a T-shirt.

the direction of the blood spatter pattern on the pants will indicate that the blood came from the floor.

It is not difficult to establish satellite spatter on a horizontal surface because the parent stain will be alongside the spatter. The difficulty arises with vertical surfaces, such as a shoe, where the item with the spatter has been moved away from the parent staining.

Satellite spatters on shoes and trousers can be misinterpreted as impact spatters from a beating or shooting and/or expirated bloodstains. Rather, they may have resulted from an injured person who has ambulated while bleeding. Factors to consider include the maximum height of spatter on an accused person's clothing, whether the bloodstaining is limited to the shoes and/or pant cuffs, whether the accused claims to have rendered assistance to the bleeding victim, and any evidence of dripped blood on the ground or floor.

5.1.3 Directionality

The shape of the blood spatter droplet, indicating the angle on which it impacted the surface, is generally more relevant to crime scenes than clothing. However, the example above of the cuffs of pants shows that occasionally the direction of travel of the droplet is important. Shoes, socks, and trousers in "stomping" cases may contain vital bloodstain pattern evidence (Ristenblatt and Shaler, 1995). The narrow end of an elongated bloodstain usually points in the direction of travel. Droplets of blood traveling up the legs of trousers indicate that the blood originated from below the legs, such as the feet or the floor. Droplets of blood on all the surfaces of a shoe, and on socks covered by trousers, may indicate the suspect was kicking and stomping on the victim rather than merely attending to a victim's injuries.

5.1.4 Clotted Blood

Clotted blood on clothing usually shows a dense center and a lighter periphery. Occasionally events take place after blood has been shed and begun the clotting process. Clotted blood observed on clothing associated with a beating death may indicate a significant time interval between blows or postmortem infliction of injury. Clots of blood may show drag patterns that indicate movement or further injury occurred after a significant interval had elapsed from the initial bloodshed. Coughing or exhalation of clotted blood by a victim may be associated with post-injury survival time.

5.1.5 Transfer Bloodstain Patterns and Contact Bloodstains

A blood transfer pattern occurs when an object wet with blood comes into contact with an unstained object or secondary surface, such as clothing. These patterns may indicate to the examiner the object that made the pattern, such as hair, a knife, or even a bloody fingerprint. A recognizable mirror image of the original surface of the object or a portion of that object may be produced. A possible transfer wipe of a bloody knife and possible contact patterns of bloody fingers on a T-shirt, among impact spatter stains, can be seen in Figure 5.6.

A contact stain is generally used as a term for a nondescript transfer stain that has been produced by contact with a bloody object but with no recognizable pattern. Figure 5.7 depicts the cuff of a woven shirt with contact bloodstains.

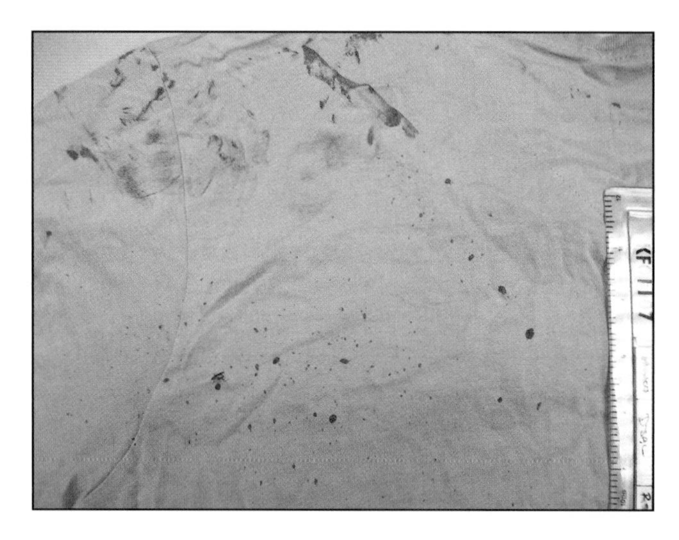

Figure 5.6 Transfer and contact patterns of a bloody knife wiped and contacted on a T-shirt.

Differentiation between a transfer pattern and an impact spatter pattern on a suspect's clothing may determine whether that suspect could have been the offender or merely someone who came into contact with the blood source. Droplets of blood lodged deep within the weave or knit may indicate spatter rather than transfer. It has been shown that medium-velocity impact blood spatter stains immediately become incorporated into the matrix of the fabric and that the stains do not remain on the surface of the fabric while drying (Miller et al., 2007). The exceptions are strongly water-repellent fabrics; for example, polyester twill used in some work pants. Blood droplets may bead up on these fabrics and collapse rather than soak in upon drying.

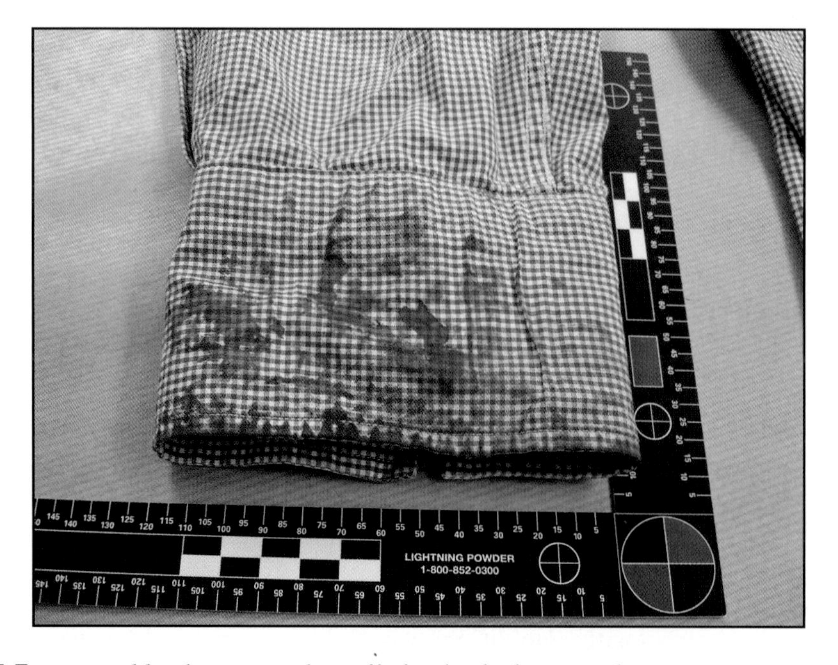

Figure 5.7 Contact bloodstains on the cuff of a checked woven shirt.

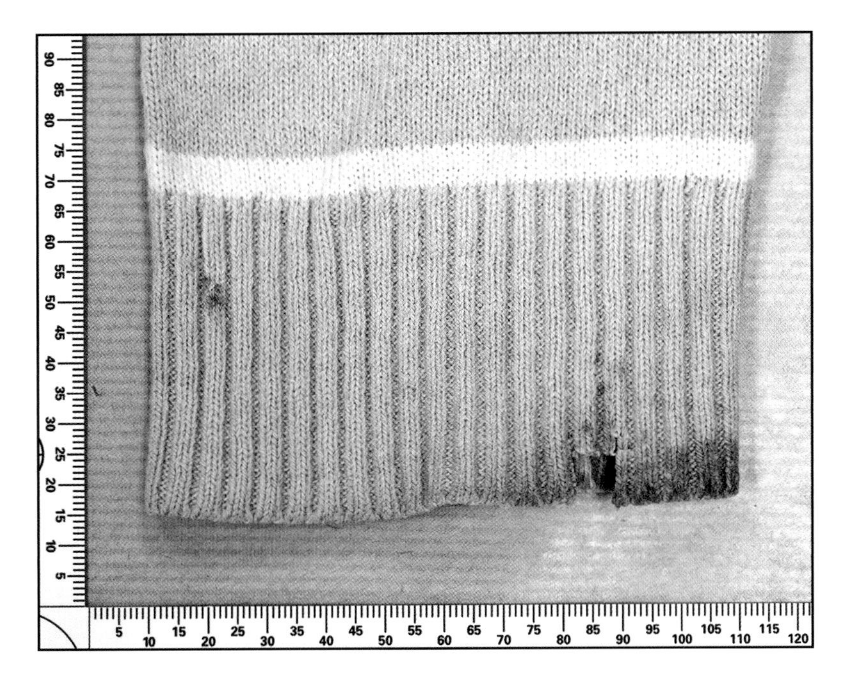

Figure 5.8 Soaking contact bloodstains on the right cuff of a sweater.

Sometimes contact bloodstains contain more information than traditionally accepted, especially when hypotheses as to their origin are tested. A suspect was apprehended for the violent rape of an elderly woman in her home in England (case study from J.M. Taupin). The victim's vagina was penetrated with a knife. Although she blocked out much of the incident, she could remember that the offender was wearing a striped sweater. Such a sweater was located in the dwelling of the suspect, and blood was seen on the garment by police. Once in the laboratory, contact bloodstains were noted on the cuffs of the sweater. A bloodstain on the right cuff appeared to have been soaked from the edges of the cuff — this accorded with the proposition that the offender had used the knife in his right hand and the victim had bled onto the hand and knife area (Figure 5.8). Bloodstains on the left cuff were contact bloodstains that appeared to have occurred from light to moderate pressure, because only the raised ribs of the knit of the cuff were bloodstained (Figure 5.9). DNA analysis of both areas of bloodstaining to the cuffs matched the subsequent profiles of the victim's blood; sections of material cut out from the garment for DNA profiling can be observed in both photographs.

5.1.6 Altered Bloodstain Patterns

When the center of a dried bloodstain flakes away and leaves a visible outer rim, it is called a "skeletonized" stain. The central area of a partially dried bloodstain altered by contact or a wiping motion may also show directionality.

Sometimes on examination, clothing may appear to have a "void" area. A void is an absence of bloodstaining in an otherwise continuous bloodstained area. The void indicates that the area was protected during the blood-letting event. The garment may have been folded during wear, or a portion of the body may have protected the area, such as sitting and shielding the waist area. Figure 5.10 depicts a running shoe with spatter and contact

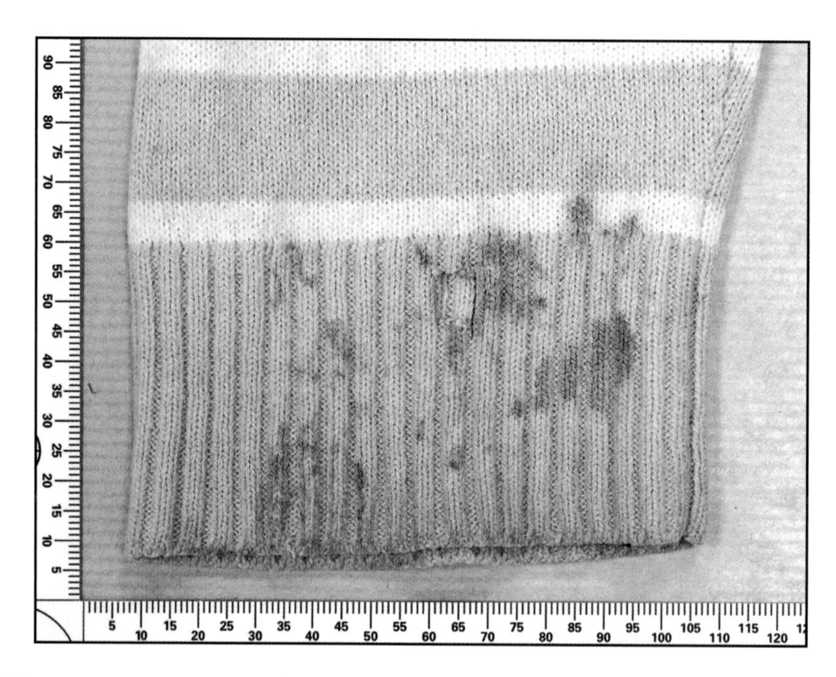

Figure 5.9 Low-pressure contact bloodstains on the left cuff of a sweater.

bloodstains. The void area corresponds to where the hem of a pair of trousers would protect the shoe from staining.

Diluted bloodstains on clothing occur in a variety of ways. Excessive moisture in the environment such as rain or snow may dilute bloodstains. Body fluids such as perspiration and urine may also dilute blood. The assailant may have attempted to clean or obliterate bloodstains, generally by washing or other addition of water. Characteristics of the original stains may be difficult to determine in these cases. A feature of diluted bloodstains is a prominent dark outer rim with a lighter center area. Cleaning methods may not completely obliterate the presence of bloodstains. Blood can be successfully removed from clothing

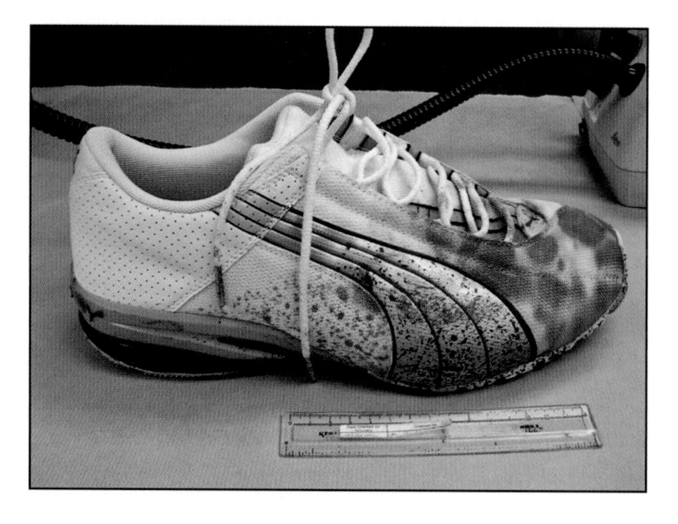

Figure 5.10 Void area on a running shoe that would have been protected by the hem of a trouser leg during a blood-letting event.

by washing, but it depends on the age of the bloodstain, the quantity of blood, the type of fabric, and the washing procedure.

Insect activity after the blood has been deposited may cause artifacts on clothing and alter the original bloodstain pattern. Usually caused by flies, insect patterns result from insects walking through blood and leaving tracks on other surfaces. They can also result from flies regurgitating or excreting small amounts of blood. Specific shapes and lack of directionality as well as unlikely locations of the blood may denote insect activity rather than original bloodstains (Benecke and Barksdale, 2003).

5.1.7 Limitations

There may be an overlap between the size of stains produced from gunshot impact spatter, expirated blood, and beating mechanisms. Thus, it may not be possible to differentiate between two or even three different mechanisms, and this should be acknowledged in any report.

The conclusions of any bloodstain pattern analysis should be based on bloodstains that are present, not on those that are expected. There are many reasons why bloodstains may not be present on the assailant's clothing (Kish and MacDonnell, 1996). It is possible to beat, stab, or shoot someone without being spattered by blood. The site of the injury may have been covered with clothing or other material during the assault and so the amount of spatter was greatly reduced or absent. The assailant may have cleaned up or changed clothing prior to being apprehended.

5.1.8 BPA Terminology Suggested for Use in Clothing Examination

Altered bloodstain: A bloodstain or pattern with characteristics that indicate a physical change has occurred

Arterial spurting pattern: Bloodstain patterns resulting from blood exiting the body under pressure from a breached artery

Backspatter: Blood directed back toward the source of energy or force that caused the spatter

Blood clot: A gelatinous mass formed as a complex mechanism involving red cells, fibrinogen, platelets, and other clotting factors

Bloodstain: Evidence that liquid blood has come into contact with a surface

Bloodstain pattern: A characteristic grouping or distribution of bloodstains that may indicate the manner in which the pattern was deposited

Bubble rings: Rings in blood that result when blood containing air bubbles dries and retains the bubble's circular configuration as a dried outline

Cast-off pattern: A bloodstain pattern created when blood is released or thrown from a blood-bearing object in motion

Contact deposit: Stain produced when clothing comes into direct contact with a blood pool or blood source

Directionality: The direction the blood was traveling when it impacted the target surface; can usually be established from the geometric shape of the bloodstain

Drip pattern: A bloodstain pattern that results from blood dripping into blood

Expirated blood: Blood that is blown out of the nose, mouth, or a wound as a result of air pressure and/or air flow as the propelling force

Flow pattern: A change in the shape and direction of a bloodstain due to the influence of gravity or movement of the object

Forward spatter: Blood that travels in the same direction as the source of energy or force that caused the spatter

Impact pattern: Bloodstain pattern created when blood receives a blow or force resulting in the random dispersion of smaller drops of blood

Impact site: The point where force encounters a source of blood

Misting: Blood that has been reduced to a fine spray, as a result of energy or force applied to it

Passive drop (bleeding): Bloodstain drop(s) created or formed by the force of gravity alone

Perimeter stain: A bloodstain that consists of only its outer periphery, the central area having been removed by wiping or flaking after liquid blood has partially or completely dried

Projected blood pattern: A bloodstain pattern that is produced by blood released under pressure, as opposed to an impact, such as arterial spurting

Satellite spatter: Small droplets of blood that are distributed around a drop or pool of blood as a result of the blood impacting the target surface

Spatter: Blood that has been dispersed as a result of force applied to a source of blood; patterns produced are often characteristic of the nature of the forces that created them

Swipe mark: The transfer of blood from a moving source onto an unstained surface; direction of travel may be determined by the feathered edge

Target: A surface upon which blood has been deposited

Transfer/contact pattern: A bloodstain pattern created when a wet, bloody surface comes into contact with a second surface; a recognizable image of all or a portion of the original surface may be observed in the pattern

Void: An absence of stains in an otherwise continuous bloodstain pattern

Wipe pattern: A bloodstain pattern created when an object moves through an existing stain, removing and/or altering its appearance

5.2 Firearm Discharge Residue Patterns

When a firearm is discharged, burnt, unburnt, and partially burnt gunpowder plus primer residue and soot are expelled from the muzzle along with the bullet. These residues are observed as concentric deposits around the bullet entry hole and can be correlated with firing distance. Approximate muzzle-to-target firing distance can be determined even in the absence of a weapon and ammunition, and more accurate determinations result from comparison with test fires using the same weapon and same type of ammunition.

Gunshot residue (GSR) particles acquire relatively high velocity from the propellant ignition (DeForest et al., 2004). Many of the gunpowder particles will be embedded in the fabric weave and may penetrate the fabric. Beyond a muzzle-to-target distance of approximately 3 feet, gunpowder deposits and gunshot residues are usually too dispersed for a pattern to be detected. The simple presence of gunpowder particles, absent a pattern, provides no information about distance. Knowing the distance the weapon was from the victim can answer questions regarding self-defense, self-infliction, and reconstruction of events surrounding the shooting.

Primer residues ejected from the muzzle and from the sides of the revolver chamber, referred to as bullet wipe, may be observed around the bullet entrance. Cotton material takes up and retains bullet wipe well, whereas some synthetic fabrics do not (Haag, 2005). Both sides of any garment containing bullet holes should be examined for the presence of bullet wipe, to determine the direction of fire.

When the bullet and other ejected materials strike the target, the bullet makes a hole and the particulate materials are deposited in a ring around the hole. Gunpowder can be observed with the naked eye and can be distinguished with a stereomicroscope. There are many types of gunpowder, including greenish disc powder, shiny elliptical ball powder, flake gunpowder, gunpowder that looks like tiny sections of pencil lead, and so on. A contact shot has most of the gunpowder and other particulates on the margins of the bullet hole, on any layers of fabric beneath, and, at autopsy, on the skin and in the wound track. In a close-up shot, a compact ring near the bullet hole is observed and particles are often driven into the fabric by the force of the blast — they can even be detected in layers of fabric beneath the outer clothing item. When the muzzle is farther away from the target (i.e., the body and thus the item of clothing), gunpowder particles will be observed more dispersed in a larger ring around the bullet entry hole. The residues usually form a recognizable pattern up to a muzzle-to-target distance of approximately 3 feet (less than that for black powder weapons). Past that, gunpowder particles tend to drop off the muzzle and may be observed among debris. These distances are approximations only.

Different firearms and types of ammunition will exhibit variation, not only in the cone of particles, but also in the persistence of particles on a given fabric (Sweeney, 2004, 2006, 2007, 2008). If bullets are recovered and the weapon suspected of being used in the shooting is available, then test fires can be conducted at different distances using the same type of ammunition. The patterns can be compared with those observed on the evidence item. Because the target fabric affects persistence of the particles, it may also be useful to do test fires using that type of fabric as the test target.

When there is no obvious pattern from gunpowder particles, when the pattern is hard to discern because of blood or other deposits, when the interpretation is uncertain, or when the examiner needs a cross-check of the data from observation of unburnt and partially burnt gunpowder particles, two types of chemical mapping tests can be performed. One type tests for nitrites in burnt gunpowder residues, and the other tests for vaporized lead from primer residues that also exit the muzzle (Bailey et al., 2006). Both fall under the umbrella terms *gunshot residue* or *firearm discharge residue*. Primer residue is also propelled out of the cartridge cylinder and back to the shooter. It can often be detected on the hands and, by extension, the sleeve cuffs. Energy-dispersive X-ray (EDX) is suggested for mapping when lead-free ammunition is used. The recently introduced millimeter-X-ray fluorescence (m-XRF) spectrometer can record mappings of gunpowder residue containing heavy-metal-free ammunition that are not accessible by chemographic coloring tests (Berendes et al., 2006); this test also has the advantage of being nondestructive. It is also possible to sample the residues and test for the organic components in gunpowder via GC-MS (gas chromatography–mass spectrometry) or, in some cases, FTIR (Fourier transform infrared spectrometry). This can be useful when it is important to establish the type of ammunition in the absence of an identifiable bullet.

On white or light-colored clothing, the dark-gray to black residue may be visible without enhancement. The pattern may require enhancement techniques on darker or colored clothing. Infrared imaging may be useful for residue obscured by blood. Other types of imaging have been used on dark-colored clothing to successfully visualize GSR patterns and powder burn patterns (Atwater et al., 2006).

Searching the clothing of the potential shooter for gunshot residue or gunpowder particles can complement examination of swabs from the person's hands. Gunshot residue on the hands is more valuable as evidence of a recent shooting, but if the person is apprehended in the short time after any gunshot residue would have been lost, the clothing, especially sleeve cuffs and lower sleeves, may provide information. However, no studies have been reported on the persistence of GSR on clothing; thus, establishing the time of deposition can be problematic.

The finding of a small number of GSR particles on hands or clothing cannot be related to a specific incident, nor does it mean that a person was in close proximity to a gun in the last several hours. Small amounts of GSR are present in police vehicles, on police personnel, and in forensics laboratories, and contamination and transfer are known to be major issues. Because gunpowder particles and gunpowder residue on a suspect can be readily transferred from surfaces such as the seats of police vehicles, and because a person standing near a shooter may exhibit gunshot residue on his or her hands and person, such results must be used with caution and an awareness of indirect transfer. The pattern of the GSR may be the valuable evidence, rather than the presence of a small number of gunshot residue particles.

5.3 Direct Contact Impressions: Imprints and Indentations

Impression marks result when a patterned object contacts a receiving surface and leaves a negative impression. Fingers of the hand, footwear or feet, weapons, and tires may all leave impressions on clothing and become evidence in a criminal case. Such patterns can aid in reconstruction of events.

Impressions should be documented by photographs before collection. The photographs should show the location of the impression on the clothing as well as the impression itself, with a scale. The plane of the film should be parallel to the plane of the impression, to avoid distortion.

Traditionally, an *imprint* is produced when an object comes into contact with a hard surface and leaves a two-dimensional representation of itself on a floor or concrete surface in dust, blood, or other medium. An *indentation* is produced by an object being impressed into a soft receiving surface, such as sand, snow, or mud, creating a three-dimensional mark. Generally, imprints are found on clothing, although the generally pliable fabric composing clothing is not the traditional hard receiving surface of walls or floors. Indentations are less frequently found on clothing because of the difficulty of producing a three-dimensional mark.

Some imprint and indentation evidence may require enhancement to be visualized, and may often be overlooked. It is important to consider the possibility of these types of patterns on clothing. Prints in blood are a special category and are discussed under blood pattern analysis as transfer bloodstain patterns.

Impressions may show class characteristics, wear characteristics, and individualizing characteristics; the latter are rarely discernable on fabric unless the surface is leather, a plastic appliqué, a silk-screened logo, or a nametag.

5.3.1 Fingerprints

Fingerprints are thought to be excellent individualizing evidence because fingerprints are considered unique. It is traditionally accepted that no two people have the same fingerprints. The presence of a visible fingerprint on clothing, not belonging to the wearer of that clothing, is valuable evidence.

Fingerprints are defined as imprints deposited on a surface by the friction ridges on a fingertip. Fingerprints without depth are often called "residue" prints. A visible (patent) print is one that needs no enhancement or development to be clearly recognizable as a fingerprint. On clothing, such a print is often made from blood, grease, dirt, paint, or other visible material. It may be suitable for comparison with no additional processing.

Bloody fingerprints on clothing have a special value because they must have been produced at the time of bleeding and before the blood dried. If they are sufficiently clear, then no enhancement may be required for comparison. However, if the ridge characteristics are not clearly defined or if the print has little residual blood, then a blood enhancement reagent may be required. The enhancement should be done in consultation with a biologist. Although most techniques do not affect DNA profiling results, there is a potential for part of the evidence to be destroyed. The priority of the identity of the blood or a identity of the fingerprint should be determined before analysis. Chemical reagents used to develop blood latent fingerprints include amido black and eosin Y (Marchant and Tague, 2007; Wang et al., 2007).

5.3.2 Footwear

The different types of information that can be collected from shoe impressions are based on the class and individual characteristics of the shoe. The impression can be identified by the tread design on the bottom of the shoe. Factors to consider when deciding to perform a lifting or enhancing method on a two-dimensional impression on clothing include the composition of the impression (blood, soil, etc.) and the nature of the material of the garment (Bodziak, 2000).

Luminol and leucocrystal violet (LCV) are commonly used reagents to develop latent bloodstains, most particularly at crime scenes. These reagents use the catalytic properties of hemaglobin. Luminol is discussed in Chapter 7 and may be useful when the material is a dark color, although the propensity for the reagent to "run" makes its use problematic. LCV does not have this effect and tends to "fix" any stain. Furthermore, light-colored clothing may be more suitable for LCV, because LCV stains latent blood a dark purple to black, allowing for easy observation. Figure 5.11(Top) shows a barely visible shoe print in blood on a white pillowcase for a case in which the offender stepped in pools of blood and then on the pillowcase on the way out of the murder scene. Figure 5.11(Bottom) depicts LCV treatment, and the sole pattern of the shoe can be visualized. Marks in clothing of readily visible bloodstains may best be processed with other reagents, such as amido black (Bodziak, 1996).

5.3.3 Tire Marks

Tire treads are composed of many different components, which gives a tire its design. Design elements are arranged in patterns or rows around the circumference of the tire and are commonly separated from one another by raised areas called grooves. Tire impressions

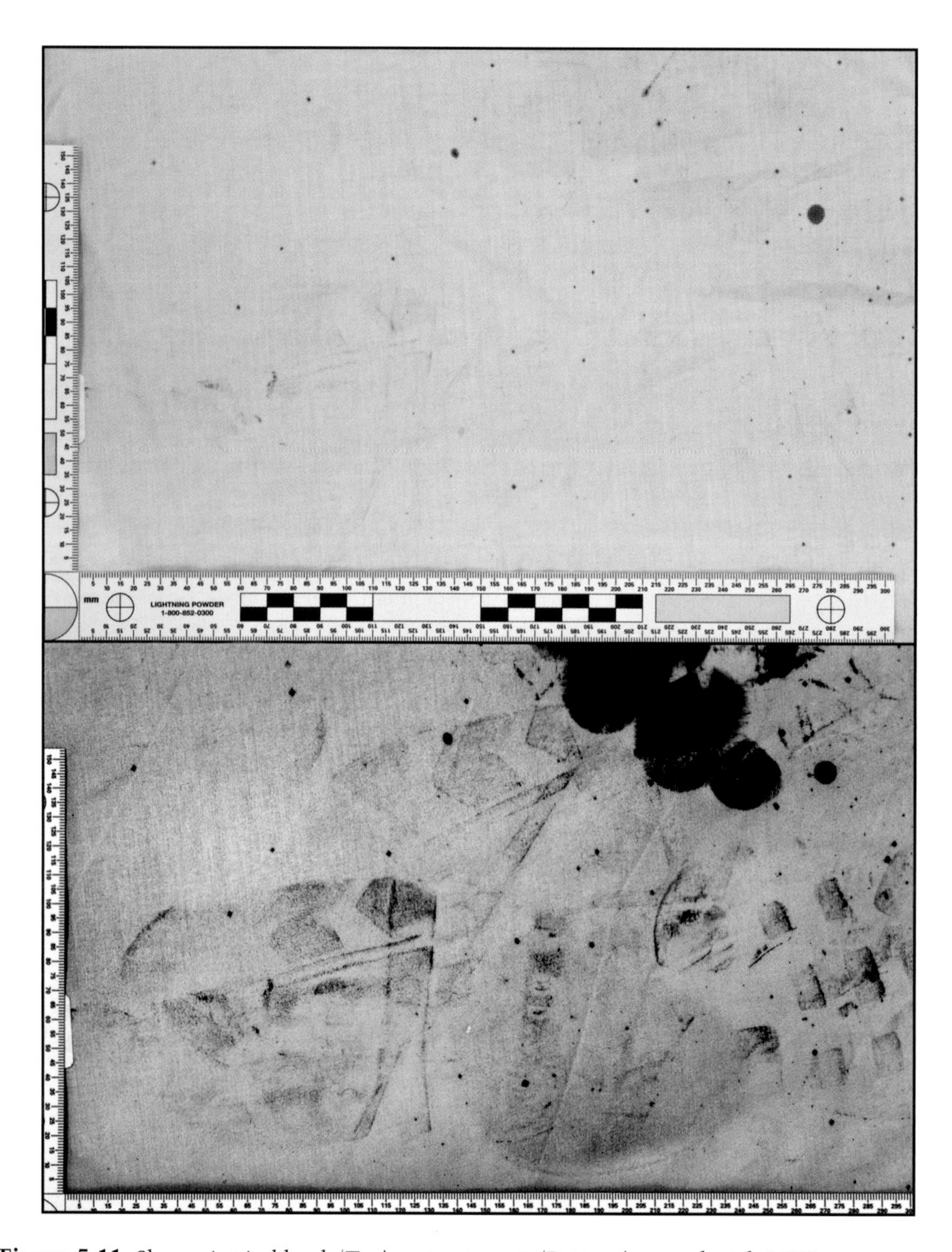

Figure 5.11 Shoe print in blood: (Top) no treatment; (Bottom) treated with LCV.

may reflect the tread design and dimensional features of the individual tires on a vehicle. Tire tread impressions can be compared directly with the tread design and dimensions of the tires from a suspect vehicle (Bodziak, 2005).

Although relatively uncommon, tire prints may be found on clothing (mainly in hit-and-run motor vehicle cases). The tire abandons the residual medium on which it has been passed, such as dirt or even blood, in the form of a print. Residue prints can be lifted with adhesive paper in much the same way as fingerprints.

CASE EXAMPLE 5.1

Homicidal injuries to midsection from automobile; tire tracks and damage to clothing

The body of a young man was found in a clearing in a wooded area. He died from the rupture of internal organs, including the spleen and liver. There were no marks on his body. Numerous tire tracks were observed in the clearing, including tire tracks that went back and forth not far from the resting place of the deceased. In the laboratory, the examiner found tire prints on the part of the victim's shirt that would have been over his midsection (exposed in scene photos). The weight of a truck on the midsection would have easily ruptured the organs in question. The police investigation led to two former high school classmates. Debris on the victim's clothing provided a strong link with the bed of the classmates' pickup. The former classmates were charged with murder, in part because the tire tracks on the clothing and surrounding area indicated that the truck ran over the victim several times, which was difficult to explain as accidental.

5.3.4 Lipstick Prints

Lipstick may indeed transfer a "lip" print to clothing. "Lipstick on the collar" is not just a phrase; it really does occur in personal contact events. Lipstick prints and smears may indicate a relationship or contact between individuals or between an individual and a crime scene. Laboratory testing can determine the chemical composition of a lipstick and even potentially extract DNA, just from the print.

Recent study of lipstick prints has shown that the quality of lip development is better on fabric than on tissue paper (Castello et al., 2002). Permanent lipsticks do not usually leave visible prints and may be overlooked, but identifiable lip prints from these lipsticks have been obtained up to 30 days after being produced (Alvarez et al., 2000). The recording of lip prints is problematic because of their highly deformable nature. The scientific admissibility of lip marks is a matter of debate in the courts and the forensic community.

5.3.5 Weapon, Tool, and Object Marks

Patterns or marks from weapons, tools, or other objects can be produced by direct contact of two surfaces. Impressions in clothing from weapons or other objects are conveyed through a medium such as soil, paint, or blood. Indentations in clothing, such as button prints or belt buckle prints on trousers or dresses, may be caused by prolonged pressure on the garment.

5.3.6 Fabric Impressions

Fabric impressions can occur in hit-and-run motor vehicle cases, where the imprint of the victim's clothing may be found on the vehicle (Kuppuswamy and Ponnuswamy, 2000). The evidence becomes even more valuable when fibers, yarns, or sections of material are transferred to the vehicle. Fabric impressions can also be observed in incidents when one item of clothing is in contact with another in an area with transferable stains or deposits.

An interesting example of transfer of fabric impressions may be shown by screen-printed clothing. Particles dislodged from screen-printed garments may exhibit fabric impressions (Amick and Beheim, 2002). The ink from the screen print can also vary in color and composition.

5.4 Physical Fit

Constructing a physical fit or a physical match in clothing examination involves fitting together sections of a garment to form either a complete garment or part of a garment. A physical fit may show that the pieces were unequivocally part of the same item. If the garment has been cut, then the physical fit will be "direct" and the pieces will fit neatly along their severed edges. For example, a sleeve cut from a wind jacket and fashioned to form a balaclava may be physically fitted to the remains of the garment if found in a suspect's house. If the edges of the pieces are torn, then the physical fit will be "indirect" and the pieces will not match neatly due to the uneven yarn ends.

5.5 References

Alvarez, S.M., Miquel, F.M., Castello, P.A., et al., Persistent lipsticks and their lip prints: New hidden evidence at the crime scene, *Forensic Sci. Int.*, 112(1), 41–47, 2000.

Amick, J.F. and Beheim, C.W., Screen-printing ink transfer in a sexual assault case, *J. Forensic Sci.*, 47(3), 619–624, 2002.

Atwater, C.S., Durina, M.E., Durina, J.P., and Blackledge, R.D., Visualization of gunshot residue patterns on dark clothing, *J. Forensic Sci.*, 51(5), 1091–1095, 2006.

Bailey, J.A., Casanova, R.S., and Bufkin, K., A method for enhancing gunshot residue patterns on dark and multicoloured fabrics compared with the modified Griess test, *J. Forensic Sci.*, 51(4), 812–814, 2006.

Benecke, M. and Barksdale, L., Distinction of bloodstain patterns from fly artefacts, *Forensic Sci. Int.*, 137, 152–159, 2003.

Berendes, A., Neimke, D., Schumacher, R., et al., A versatile technique for the investigation of gunshot residue patterns on fabrics and other surfaces: m-XRF, *J. Forensic Sci.*, 51(5), 1085–1090, 2006.

Bodziak, W.J., Use of leucocrystal violet to enhance shoe prints in blood, *Forensic Sci. Int.*, 82, 45–52, 1996.

Bodziak, W.J., *Footwear Impression Evidence: Detection, Recovery, and Examination*, 2nd ed., CRC Press, Boca Raton, FL, 2000.

Bodziak, W.J., Forensic tire impression and tire track evidence, in *Forensic Science: An Introduction to Scientific and Investigative Techniques*, 2nd ed., James, S.H. and Nordby, J.J., Eds., Taylor & Francis, Boca Raton, FL, 2005.

Brown, M. and Wilson, P., *Justice and Nightmares: Successes and Failures of Forensic Science in Australia and New Zealand*, New South Wales University Press Ltd., Kensington, NSW, Australia, 1992.

Castello, A., Alvarez, M., Miquel, M., et al., Long lasting lipsticks and latent prints, *Forensic Sci. Communications*, 4(2), April 2002.

Chisum, W.J. and Turvey, B.E., *Crime Reconstruction*, Elsevier Academic Press, Burlington, MA, 2006.

DeForest, P., Martir, K., and Pizzola, P.A., Gunshot residue particle velocity and deceleration, *J. Forensic Sci.*, 49(6), 1–7, 2004.

Haag, L.C., *Shooting Incident Reconstruction*, Elsevier Academic Press, Burlington, MA, 2005.

Inman, K. and Rudin, N., *Principles and Practice of Criminalistics: The Profession of Forensic Science*, CRC Press, Boca Raton, FL, 2001.

James, S.H., Literature search for the distance that backspatter travels, *IABPA News*, September 2007.

James, S.H., Kish, P.E., and Sutton, T.P., *Principles of Bloodstain Pattern Analysis: Theory and Practice*, 3rd ed., CRC Press, Boca Raton, FL, 2005.

Karger, B., Nusse, R., and Bajanowski, T., Backspatter on the firearm and hand in experimental close-range gunshots to the head, *Am. J. Forensic Med. Path.*, 23(3), 211–213, 2002.

Kish, P.E. and MacDonnell, H.L., Absence of evidence is not evidence of absence, *J. Forensic Identification*, 46(2), 160–164, 1996.

Kuppuswamy, R. and Ponnuswamy, P.K., Note on fabric marks in motor vehicle collisions, *Sci. Justice*, 40, 45–47, 2000.

Marchant, B. and Tague, C., Developing fingerprints in blood: A comparison of several chemical techniques, *J. Forensic Identification*, 57(1), 2007.

Merchant, N.A., The McLeod-Lindsay case, *Aust. Police J.*, 21(3), 181–196, 1967.

Miller, M.T., DiMarchi, E., and Zhang, L., The microscopic characteristics of drying and transfer of impacted bloodstains on fabric and textiles, *Proceedings of the American Academy of Forensic Sciences*, Vol. xiii, February 2007. Available at http://aafs.org/pdf/07Proceedings.Complete.pdf/; accessed April 2008.

Pex, J.O. and Vaughan, C.H., Observations of high velocity bloodspatter on adjacent objects, *J. Forensic Sci.*, 32(6), 1587–1594, 1987.

Ristenbatt, R.R. and Shaler, R.C., A bloodstain pattern interpretation in a homicide case involving an apparent "stomping," *J. Forensic Sci.*, 40(1), 139–145, 1995.

Slemko, J.A., Bloodstains on fabric: The effects of droplet velocity and fabric composition, *IABPA News*, December 2003.

Sweeney, K.M., *The Interpretation of Gunpowder Particle Deposition and Impact Characteristics*, Criminalistics Section, scientific session, Presented to the American Academy of Forensic Sciences meeting, Dallas, TX, February 20, 2004.

Sweeney, K.M., *Gunpowder Stipple Patterns of Commonly Encountered Small Firearms*, Criminalistics Section, scientific session, Presented to the American Academy of Forensic Sciences, Seattle, WA, February 24, 2006.

Sweeney, K.M., *Gunpowder Particle and Vaporous Lead Deposit Patterns on Fabric from Hand Gun Discharges*, Criminalistics Section, scientific session, Presented to the American Academy of Forensic Sciences, San Antonio, TX, February 24, 2007.

Sweeney, K.M., *Gunpowder Particle and Vaporous Lead Deposit Patterns on Fabric from Hand Gun Discharges II*, Criminalistics Section, scientific session, Presented to the American Academy of Forensic Sciences, Washington, DC, February 23, 2008.

SWGSTAIN, Scientific working group on bloodstain pattern analysis: Recommended terminology, *IABPA News*, June 2008.

Teichroeb, R., Forensic scientist in Washington crime lab tied to wrongful convictions in Oregon, *Seattle Post-Intelligencer*, December 27, 2004.

Wang, Y., Zheng, W., and Ma, J., Eosin Y detection of latent blood prints, *J. Forensic Identification*, 57(1), 2007.

Damage

6

6.1 Introduction

Damage to clothing of the victim of a crime, or indeed the perpetrator, is often found in serious crimes of violence such as homicide and rape. Items of clothing are torn in a struggle, cut by knives in assaults, or have ruptures or holes created by gunshots. Clothing may also be damaged for use in other crimes, such as burglary or armed robbery — the creation of masks from sweaters, for example. Garments may even be damaged by an alleged victim in order to mimic an assault (Taupin, 2000). Damage to clothing can also reflect environmental insult, such as heat damage, insect attack, or decomposition due to time and weather.

Characteristics of damaged clothing may provide information as to the implement that may have caused the damage, the manner in which it was caused, and whether it was "recent"; damage analysis may corroborate or refute a crime scenario (Monahan and Harding, 1990; Taupin, 1998b). The more distinctive the weapon and the more outstanding the characteristics in the damaged clothing, the stronger is the link that may be achieved between the weapon and the damage. One example is that of a crossbow arrow with a broadhead tip, which produces characteristic Y-shaped cuts when fired into textiles (Taupin, 1998a). On occasion, especially when the body is badly decomposed, the damage to the clothing of a deceased person may reflect a clearer geometry of the weapon than does the associated wounding. Sometimes, the body is completely decomposed or absent, and thus any clothing obtained supplies the evidence of death or injury.

Further information as to the weapon used to cause the damage may be gained when multiple areas of damage in varying types of fabric are encountered in one case. The following attempted murder case (Taupin, 1999) illustrates not only the value of multiple damage analysis but also the value of outstanding characteristics in that damage when there is a suspect weapon:

> A 92-year-old Melbourne woman was attacked in her home by a bank teller from her local bank. The bank teller had stolen money from the elderly woman's account to fund a gambling habit and was frightened when she complained to the bank about missing funds. Although the victim was stabbed 18 times in the stomach and back, she survived the attack, no doubt due to the large number of garments she was wearing, which acted like bandages. There were a total of 49 stab-type cuts to the 7 items of clothing, predominantly to the garments on her upper body. Several of the cuts in the more tightly knitted garments displayed a distinctive Y-shaped feature at one end. These cuts were similar in profile to simulation cuts produced by the suspect weapon, a boning knife located in the home of the accused. Furthermore, this boning knife was sufficiently sharp to penetrate all the clothing layers. Although the damage evidence was not accompanied by any other biological or chemical evidence, it provided sufficient weight to link the weapon to the victim. The accused pleaded guilty at trial.

The usefulness of damage analysis, however, like other techniques already outlined in this book, initially depends on the recognition by the police investigator and the forensic scientist that it could provide valuable evidence.

In order to interpret damage to clothing, one must understand how fabrics that compose clothing are constructed. Damage is as much a property of the fabric as of the implement imparting the damage. Chapter 3 summarized fabric construction and yarn type, but the reader is encouraged to gain a personal knowledge base in this area through reading reference material and experimentation.

Damage to clothing is dominated by observation of the morphological characteristics at both the macroscopic and the microscopic level. Characteristics of clothing damage cannot be assessed in numerical terms. One of the few published studies assessing participants in a clothing damage trial used blind proficiency tests on people with varying levels of forensic awareness, experience, and training (Boland et al., 2007). The "forensic" group, comprising attendees at a forensic science conference and including a range of professionals, consistently performed better than the "student" group, comprising students of a forensic science degree course. However, both groups showed room for improvement in assessment of both fabric construction and damage identification. A systematic approach to damage maximizes the benefits and helps minimize the subjectivity in the field (Boland et al., 2007).

It is especially important with damage evidence to embrace the nonprescriptive holistic approach that we adopt in this book, coupled with an objective description of damage characteristics and the study of test damage by the examiner.

6.2 Damage Categories

The potential causes of damage to textiles can be broken down into five broad categories: mechanical effects, chemical effects, photochemical effects, microbial and other biological damage, and the influence of heat (Taupin et al., 1999). Mechanical effects, such as cutting and tearing, are the most commonly encountered in forensic casework. The following subtypes of damage may generally be differentiated according to their morphological (physical appearance) characteristics, although there may be some overlap. The most prevalent subtypes will be discussed in detail later in this chapter. Further definitions of terms are presented in the glossary at the end of this chapter.

1. *Cut.* A severance with neat edges produced by a sharp-edged instrument. Types of cut include:
   Stab cut. Typically produced by a knife. (See Figure 6.1.)
   Slash cut. Typically produced by a sharp-edged tool such as a knife, razor blade, or scalpel. (See Figure 6.2.)
   Scissor cut. Produced by the opening and closing action of scissors as they cut along material. (See Figure 6.3.)
2. *Tear.* A severance caused by the pulling apart of a material, leaving ragged or irregular edges. (See Figure 6.4.)
3. *Puncture.* Penetration through material by an implement, producing an irregular hole; generally produced by a blunt implement in a thrusting action. (See Figure 6.5.)
4. *Abrasion.* Caused by the material rubbing against another surface; may result in thinning of the material, even holes and fraying.

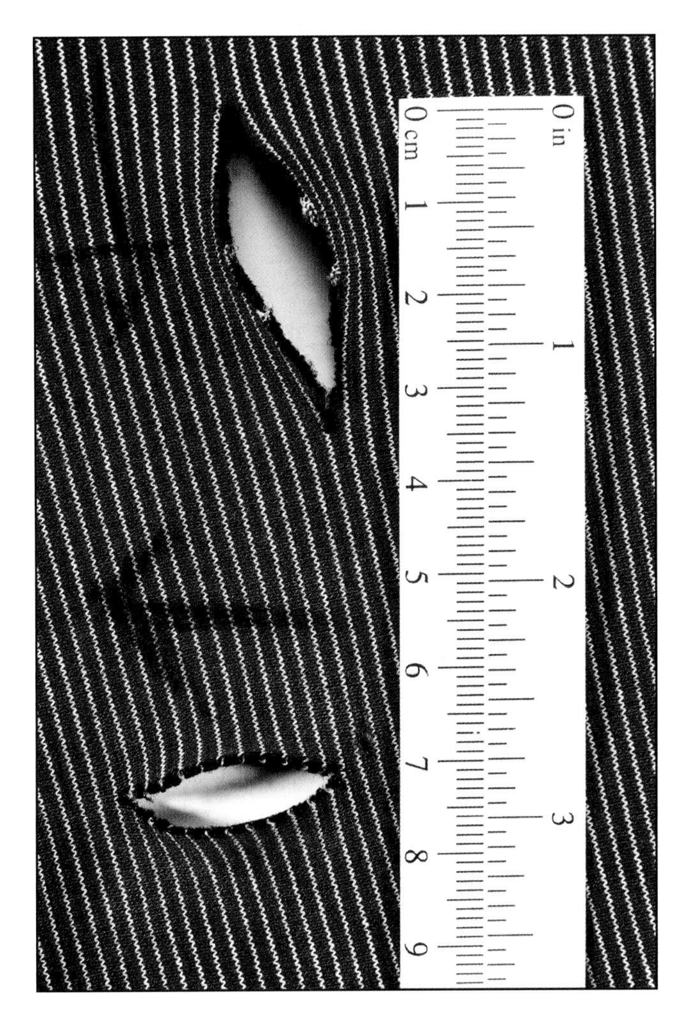

Figure 6.1 Stab-type cut in a knit.

5. *Animal damage.* Bite marks and other severances produced through the jaws and feet of an animal (e.g., canine or rat) or the beak and feet of a bird. Insects such as moths and beetles may ingest the fibers of the material, producing small, puncture-like holes.
6. *Microbial damage.* Irregular damage to material, most often seen in burials in soil. Microbes such as bacteria and fungi may destroy fibers. This type of damage preferentially affects natural fibers, on occasion leaving synthetic material unaffected. A mix of fibers in a garment may lead to preferential damage; "pseudo" cuts may even be produced.
7. *Thermal damage.* Damage to the material may range from minor, such as slight scorching, to complete combustion.
8. *Tensile failure.* Fracture of the material through pressure, especially in ropes and webbing; may be found in suicide by hanging, boating accidents, etc. Determining the force required to produce tensile failure requires special knowledge and

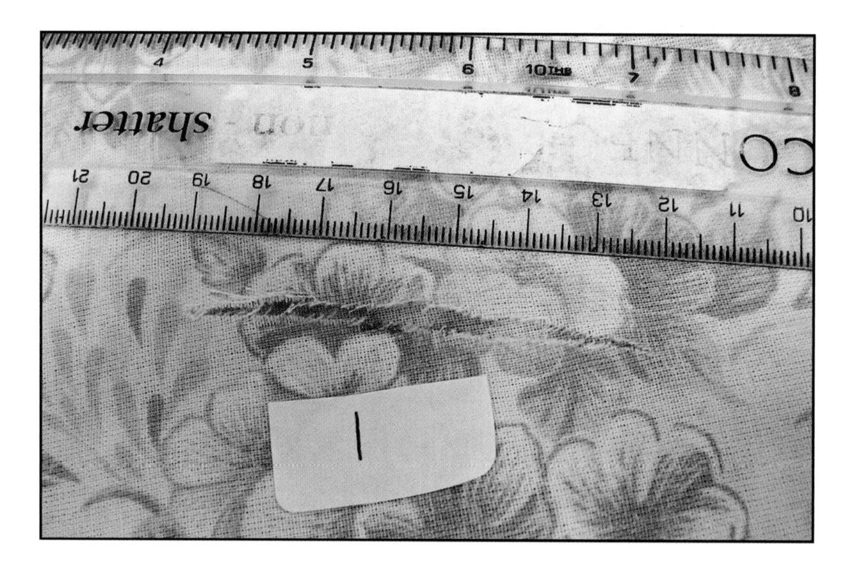

Figure 6.2 Slash-type cut in a weave.

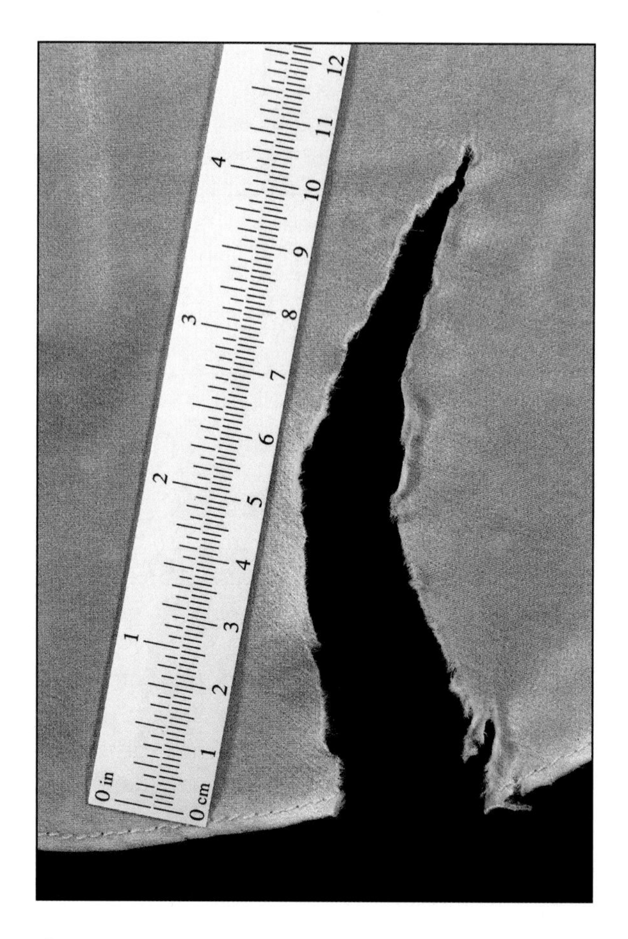

Figure 6.3 Scissor cut.

Figure 6.4 Tear caused by pulling apart of material.

instruments and may require consultation with an expert with the relevant background and equipment. Fracture mechanisms and single fiber failure have been described and illustrated in an atlas (Hearle et al., 1998).

Combinations of two or more of any of the above are sometimes seen.

6.3 Examination Approach

When determining the examination method, the scientist should consider the results of previous research, the principle that damage characteristics will reflect their cause, and the use of simulation experiments (Taupin, 1996). We suggest a three-level approach to the examination of damage to clothing (Johnson, 1991; Taupin, 1996; Taupin et al., 1999):

1. Fabric
   ↓
2. Yarn
   ↓
3. Fiber

Such an approach gains the maximum level of information.

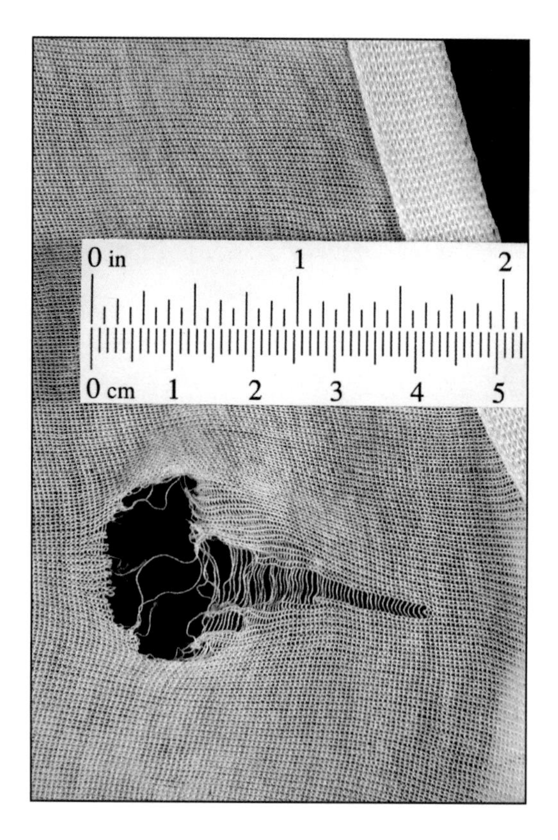

Figure 6.5 Puncture.

Studies in the literature (Stowell and Card, 1990) have used fiber end appearance as an aid in determining the cause of textile damage, using scanning electron microscopy (SEM). However, doubts have been raised about both the validity and reliability of the SEM technique in textile damage interpretation (Morling, 1987; Upkabi and Pelton, 1995). Fiber end morphology alone may be unreliable to distinguish the cause of fiber damage (Pelton, 1995). Manufactured fibers may be produced as staple or multifilament yarns (see Chapter 3). The method of tow conversion (cut or stretch broken) determines the fiber end appearance of the "constituent" staple fibers (Pelton, 1995). If the clothing fabric has been made from staple fibers and the damaged edges are difficult to view, the fiber ends observed might represent a distribution of both constituent and damaged fiber ends instead of damaged fibers alone. Fiber end features may also be attributed to or influenced by various manufacturing processes (Pelton, 1995).

Consequently, only on specific occasions is it suggested that SEM be used in clothing damage analysis (such as thermal damage to fibers).

The item of clothing to be examined should first be classified according to its construction (see Chapter 3). The damaged area should then be examined optically with the naked eye and then a stereomicroscope with variable magnification to at least 40 times. The use of a stereomicroscope with a discussion head is of considerable advantage because it allows two damage examiners to view the same features simultaneously. The following levels should be examined:

1. *Fabric level.* The location of the damage on the garment, the extent, and the profile should be noted. Severance lengths should be documented. When a case has multiple garments, it is useful to use a mannequin to correlate any damage. If multiple layers of fabric have been damaged from the clothing of one person, a comparison of the various fabrics may be informative, especially if they are of different fabric types. The examiner should be aware of the possibility of damage through folds of material. Distortion of the fabric around any severance should be noted, such as buckling or folds out of the fabric plane and changes to the normal thread spacing, including runs in knitted fabrics. Direction of the severance line relative to the thread direction and relative positions of the severed yarn ends may also be important.
2. *Yarn level.* The severed ends of the yarns themselves (neat or frayed) should be noted. Features such as planar array, nicks, and steps should be observed. Snippets in knit fabrics should be collected. The end areas of a severance should be noted, especially where partly cut yarns are observed.
3. *Fiber level.* This level should be noted but generally provides less information. Fiber ends from white or pastel fabrics may be difficult to view under an optical microscope, so differential staining of fiber types can make this easier. Stains such as those in the Shirlastain range or special lighting may assist.

Figure 6.6 illustrates these levels of examination.

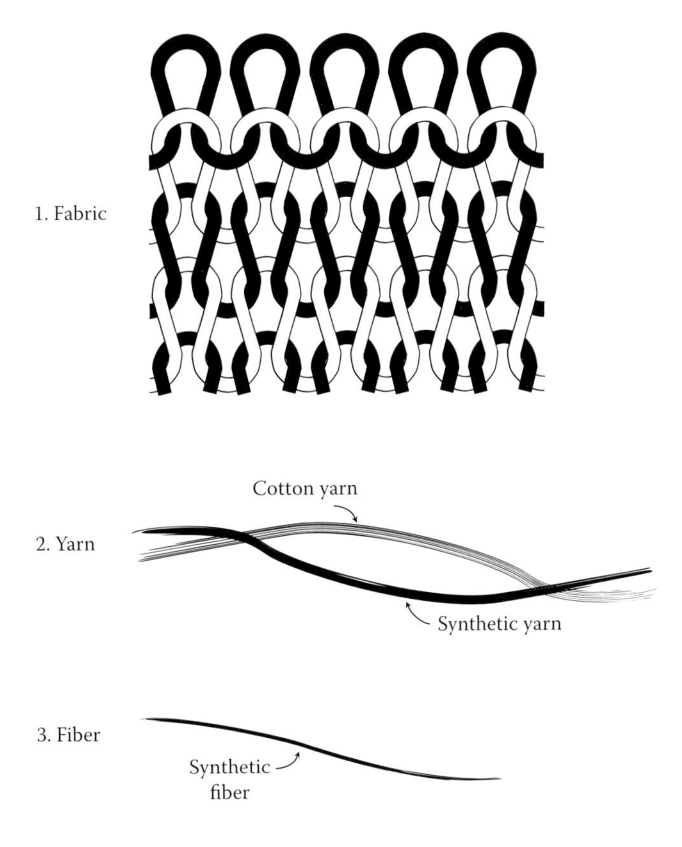

Figure 6.6 Levels of examination.

Fiber, yarn, and fabric textile variables interact with one another. An investigation involving alleged cuts or tears in a garment should thus also involve the ability of the material of the garment to tear. The severance direction (warpwise, weftwise, or diagonally across the fabric) should always be noted. Simulation experiments (see Section 6.9), in which the fabric is damaged according to a proposed scenario, can also be invaluable.

It is especially important to observe areas of body fluid soiling and its position in relation to any damage. Blood and other soiling may "tidy up" severed edges, complicating interpretation. Before performing any simulation experiments, you will need to remove evidence to be tested, such as blood and fibers.

6.4 Normal Wear and Tear and "Recency"

It is necessary to first determine whether the damage examined may be related to a crime, or whether it has been produced on the garment through normal events or at some previous time. These conclusions are drawn through the observation of the damage and associated matter. However, in extreme cases, such as the damaged garment being stored in a drawer for years, no alteration to the garment would be observed. Conversely, the wearing of the garment since its damage or subjecting the garment to environmental insult (such as burial) after damage may alter this damage.It is thus imperative to obtain the history of the clothing items before examination.

Damage to clothing of interest in investigating falls, pushing, dragging, and tearing or cutting of fabric must be distinguishable from normal wear and tear. Most garments, even if they have just been purchased, show some signs of alteration since manufacture. Even placing a fragile knitted garment on a hanger in a clothing store may cause the garment to lose shape or produce "runs" in the knit. It does not surprise (although may disappoint) purchasers of "new" garments from retail stores that such garments may have faults such as loose buttons, pilling, and fraying of cuffs and hems. These features occur due to the exposure of the garment to the environment and the public. More extensive wear and tear, such as that produced by years of wearing and washing, may include unraveling of hems and seams, snags (especially in stockings), thinning of fabric, and holes. Old tears may be noted, and the repair of this normal wear and tear, such as darning, can be readily ascertained. Figure 6.7 shows thinning of material in the heel area of a knitted sock. Wear and tear also includes damage produced in the course of work. Examples include burn holes on the clothing of a welder and abrasion accompanied by grease deposits on the knees of a mechanic's pants.

Although it may not be certain that such damage was produced during the incident under investigation, signs of buttons having been ripped from a shirt and "fresh" tears and cuts or punctures are *not* considered normal wear and tear. The concept of whether the damage is over and above normal wear and tear is often accompanied by that of "recency." The determination of recency is often made by the observation of additional features of the damage. This is generally the first issue to be addressed in damage interpretation, because nonrecent damage is usually not relevant to the incident under investigation and, other than being noted, need not be pursued. Most damage that is not recent has features that are produced by prolonged wear or by washing. Cut yarns have fibers that begin to fray on wear since the damage. Shorter, extraneous fibers from yarns out of the plane of the material also appear. Once a garment is washed, the fibers on the severed edges of the damage mat together and become tangled. Sometimes, if the garment is washed with other

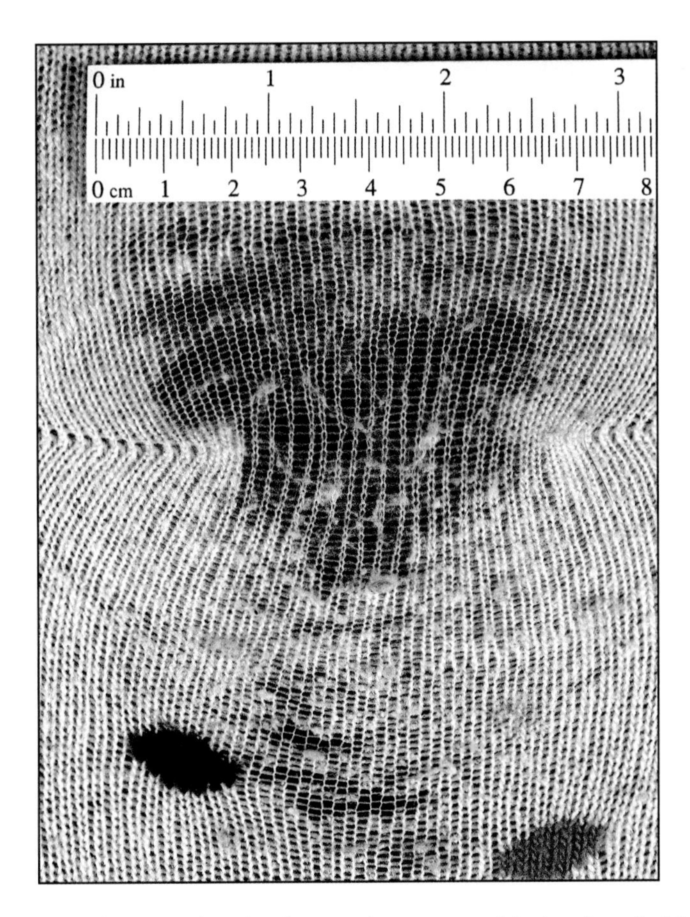

Figure 6.7 Thinning of material in heel area of a patterned knitted sock. Note alteration in yarn spacing.

clothing, fibers from the other garments may become caught in the severed edges of the damage. Consequently, the presence of foreign fibers along the severed edges of a damaged garment, especially if the severed edges are matted and tangled, indicates that the damage is not recent.

Other indications of potentially recent damage include the garment being in a condition that makes it impossible for it to be worn in the normal fashion. For example, broken or missing hooks and eyes in brassieres, broken zippers, broken or missing press studs or buttons, and severed brassiere straps would be considered recent.

6.5 Cuts

Cuts in a garment are characterized by their neatness and their ability to change direction without significant distortion to the fabric. Yarn ends in a cut will usually line up quite well. This feature is emphasized when looking at the yarn level itself. The fiber ends in a cut yarn will usually be in the same plane; this has been referred to as "planar array" (Morling, 1987; Taupin et al., 1999). Short, separated segments of yarn may also be created in a cut. For example, cuts in knitted fabrics may produce "snippets" if the fabric is cut at an angle to the yarn direction (Figure 6.8).

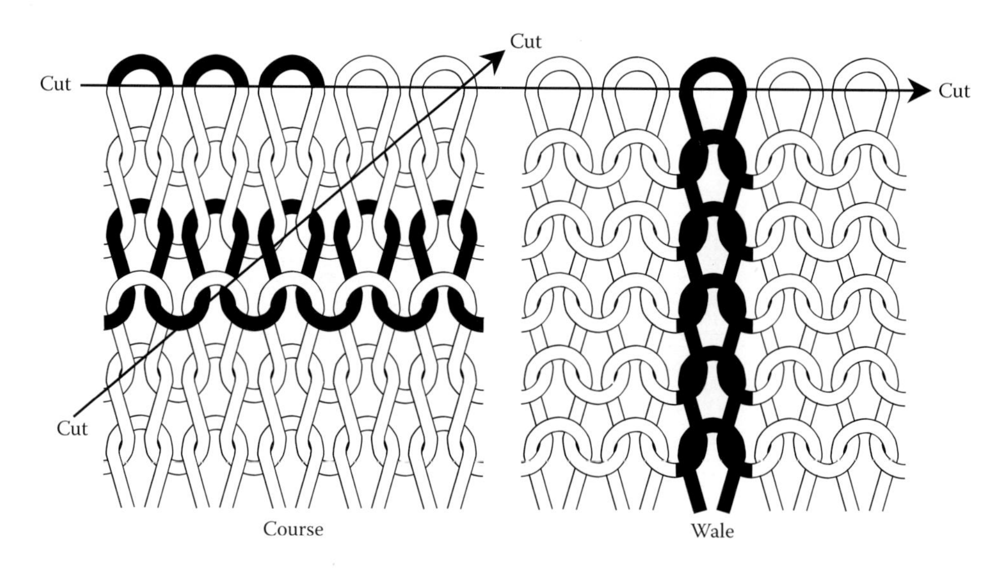

Figure 6.8 Snippets produced in cutting.

Cuts to clothing may be broadly categorized into stab cuts, slash cuts, and scissor cuts. Stab cuts are penetration of the clothing by a cutting implement in a thrusting manner; most cuts in criminal cases are these, and a section below describes them in more detail. Slash cuts are cuts produced in clothing in a sweeping or "slashing" manner and may not involve penetration. A slash cut may start and finish in a V shape and may not be continuous (Figure 6.9a, b). Sometimes only the surface yarns will be cut in a slash cut.

Scissor cuts are a special type of clothing damage, because scissors are generally used for a specific purpose in clothing construction, such as cutting out pieces of material. Medical staff may use scissors to quickly remove clothing from an injured person (hospital-type damage). Sometimes, scissors are used in false reports to mimic assault (Taupin, 2000). Offenders have also used scissors to cut the clothing of their victims postmortem. The actions of the opening and closing of the two blades of the scissors produce characteristics that may be able to identify their use. The presence of scissor-cut "stoppages" indicates the areas where the scissor blades closed and opened to commence a new cut. "Tongues" of cut material also indicate the use of scissor blades; rough scissor cutting or scissor cuts through folds may produce this feature. Abrupt changes in direction of the cut, especially more than one change, are strongly indicative of scissor damage (see Figure 6.10 for examples).

Cutting indicators, depending on material:

In a weave

No preferred direction — an ability to track in any direction, a change in direction, etc.
Relatively featureless edges
An ability to fiber-end or pattern match
Presence of a significant planar array

Figures 6.11 shows a cut and tear in a weave. The cut commenced the action through the hem of the dress, and this severance was extended by manual pulling apart to produce a tear.

In a knit

No preferred direction — an ability to track in any direction
A change in direction

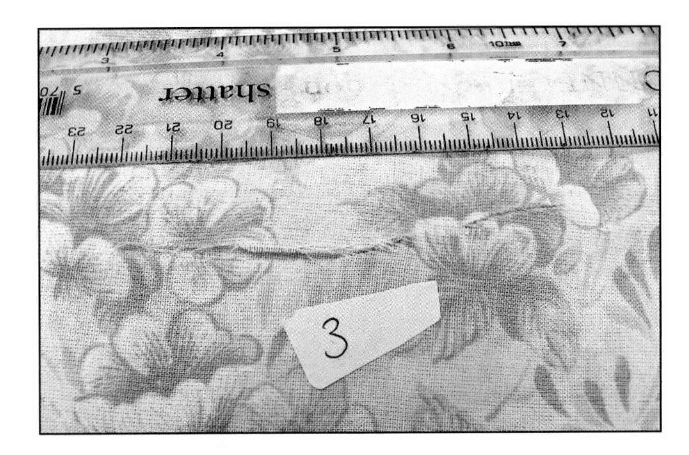

Figure 6.9a Slash cut showing discontinuity.

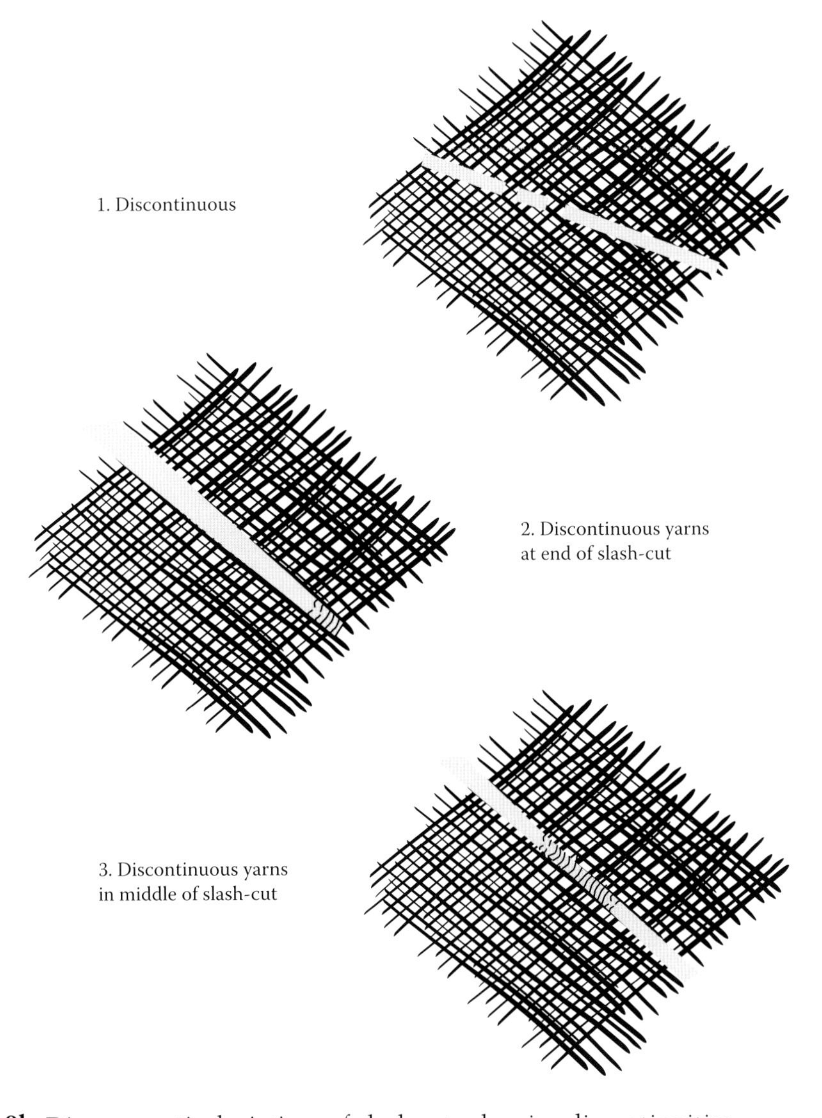

Figure 6.9b Diagrammatic depictions of slash cuts showing discontinuities.

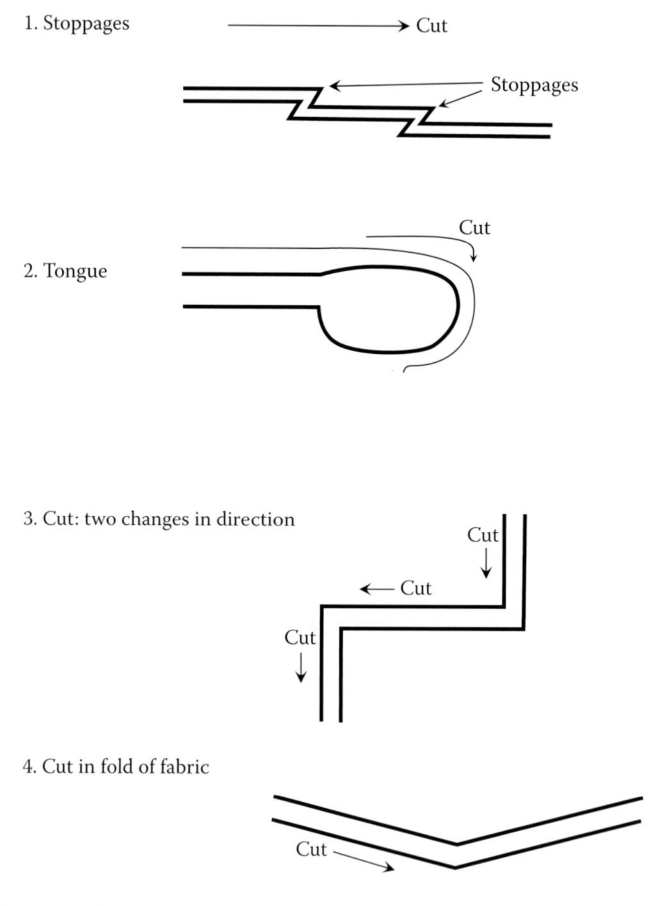

Figure 6.10 Indicators of scissor cuts.

Discontinuities typical of scissor-cut "stoppages"
Presence of tightly bound "tufts" or "snippets" along the severed edge
Presence of a significant planar array

In non-woven materials (felt) and leather — Edge lines are neat

6.6 Tears

Tears in garments often occur through manual pulling apart of the material. It is also possible to tear a garment by applying force to a defect already present, such as extending a cut by tearing. Snagging a piece of material on a nail or other penetrating object may cause a tear in the garment by extending the existing puncture through movement of the body.

Tearing is characterized by distortion and rupturing of the clothing. Tears in a garment usually propagate parallel to one of the thread directions. A given textile fabric could tear in both directions (warp and weft), in only one direction (warp or weft), or in neither direction (only distort the yarn). Tearing will cause the fibers to break at different positions along the yarn, leaving a less well-ordered yarn end. Fraying of the yarn ends will also occur.

It is very difficult to commence a tear "from scratch." Most tears will initiate from either a weak point in the garment or from another defect such as a cut or a hole. However,

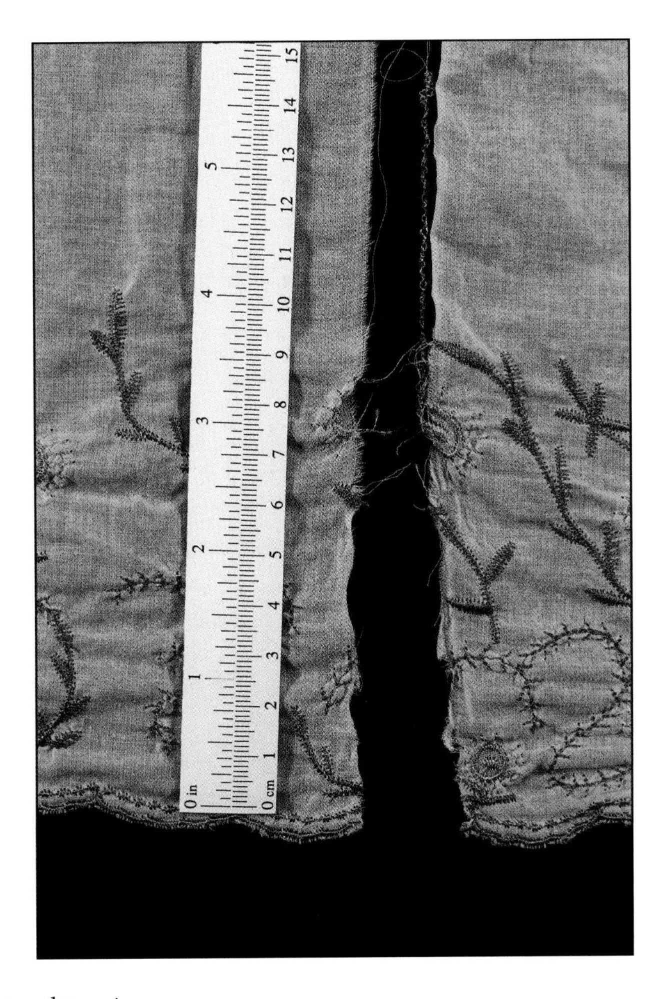

Figure 6.11 Cut and tear in a weave.

it is possible to tear a garment without initiation from a weak point if the garment material is flimsy, or if the force is great such as the undercarriage of a vehicle. Simulation experiments will support any theories. Figure 6.12 illustrates a tear that has commenced from the top of the back zipper in a skirt. The material in this skirt tore like paper.

Tearing indicators in a weave
Damage follows preferred direction of tear, parallel to the warp or weft
Associated stretching
Edges devoid of "planar array"
Noticeable "curling over" of the fabric along the severance line

Tearing indicators in a knit
Damage follows preferred direction of tear
Associated stretching
Distortion of fabric along the severance line
Noticeable "curling over" of the fabric along the severance line
Generally ragged and discontinuous edges

Non-woven materials (felt) and leather – edge lines are clearly ragged or fibrous.

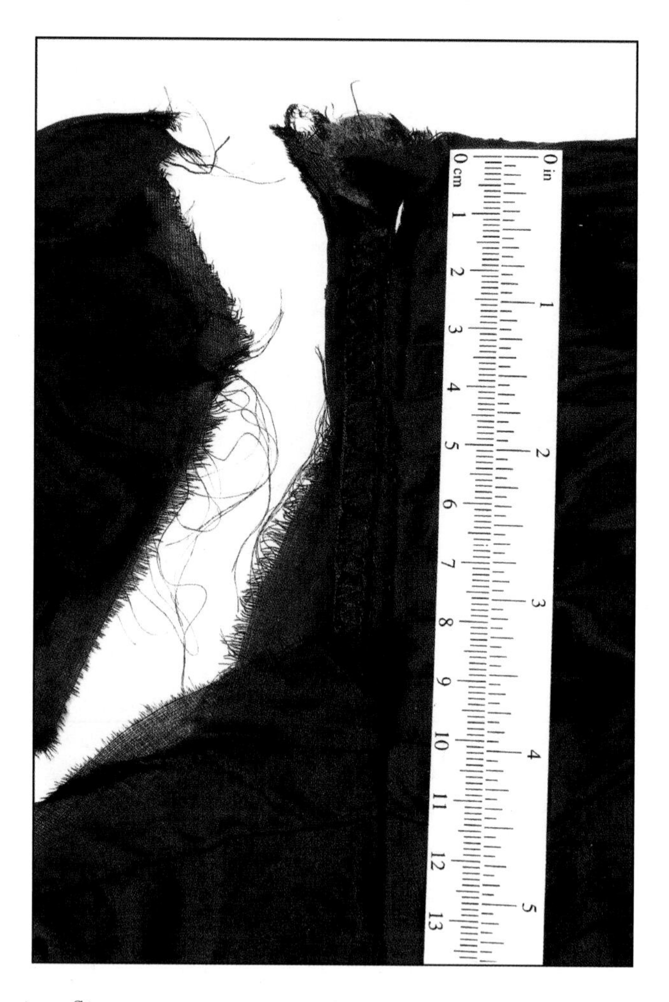

Figure 6.12 Tear in a flimsy weave.

6.7 Holes and Punctures

A hole is defined as a defect in a garment, most often roughly circular in shape, and most often acquired in clothing through wear and tear. Thin or fragile garments, or clothing washed and worn extensively, are most susceptible to this type of damage — for example, stockings, fine knitted tops, and undergarments.

Punctures are produced in stabbing actions resulting from a blunt implement, such as a screwdriver. The yarns are stretched and break unevenly, leaving a square or oval hole. The damaged area appears rough and it is difficult to see where the broken yarns had joined; consequently, features of tearing are observed. A closed pair of scissors, when used in a stabbing action, will produce a puncture, especially if the tips are rounded.

6.8 Stabbing

Stabbing with an implement (most commonly a knife) may incorporate cuts, tears, and punctures. Stab-type injuries and fatalities are reported to be one of the most common crimes of violence in Britain (Hunt and Cowling, 1991; Rouse, 1994), Europe, and Australia.

Those countries have banned firearms without a special permit, and consequently they have fewer firearms in circulation than in the United States (see later for firearm damage).

A Swedish study (Ormstad et al., 1986) showed that in about 62% of sharp force fatalities, a knife had been picked up at the site of the killings. These fatalities occurred indoors in a domestic environment and the knives were of a household type (carving knife, bread knife, paper knife). In the remaining cases, either the perpetrator or the victim brought a knife or sharp tool to attack or defend, and among these, sports knives, work knives, or clasp knives dominated. Screwdrivers, scissors, stilettos, bayonets, and serving forks were used in one or two cases each.

Both anecdotally and in the literature there has been a rise in what is termed "knife crime" in the United Kingdom. Between 1997 and 2005, the number of people admitted to the hospital in England following assault with a sharp object rose by 30% (Maxwell et al., 2007).

The chest is the most common site for both single and multiple fatal stab wounds, and it would be expected that the chest would be clothed in most cases. Thus, the clothing may provide important evidence, especially if the body is absent or reporting is not timely.

The dynamics of stab wounds has traditionally involved stabbing tests in the flesh of corpses. Knight (1975) found that ease of penetration depended on the cross-sectional area of the knife tip; skin was the most important tissue encountered, and once a weapon penetrated the skin it continued with very little force. The sharpness of the tip of the weapon was paramount in all considerations of the amount of force needed to cause any stab wound. Green (1978) performed experiments on fully clothed cadavers with a variety of knives and found that although greater force was involved in penetrating the clothing plus skin, the most force was needed to remove the knife from the body. Rocking or twisting was frequently needed to remove the larger knives from clothed bodies.

Some generalizations can be made regarding stab cuts in clothing produced by knives. The blunter the tip of the knife (that is, the larger the cross-section of the tip), the greater is the distortion produced in the fabric. Stab-cut dimensions in clothing do not accurately reflect the knife blade width (Costello and Lawton, 1990). Variables such as the degree of stretch of the fabric, whether the garments are taut or loose, and the angle of the blade to the fabric all affect the length of a stab cut.

The sharpness of the blade will affect the shape of the severance. A sharp blade will produce neatly cut yarns, in contrast to the "beard" pointed yarn ends seen in a tear. A blunter blade will tend to pull the yarns before eventually cutting them, resulting in distortion and increased fraying. It is possible to have combination cuts and tears produced in a stab by a blunt knife that still may cut yarns or fibers.

Features such as serrated edges on the blade and notches may produce distinctive characteristics along the severed edges. It is occasionally possible to match defects along the edge of the blade with defects in the fabric.

If one end of the cut is neat but the other has frayed fibers and distorted yarns, then a knife with a single smooth blade is indicated, or a double-bladed knife with one serrated and one smooth edge. If a knife is provided as a suspect weapon, it is often possible to determine the "directionality" of a stab cut; that is, the orientation of the blade to the fabric. Single-edged blades with a thick, blunt back edge produce a tapered severance, because the blunt back pushes yarns away from the severance near the point of initial penetration (Figure 6.13). Simulation experiments (see next section) with the suspect knife may be required.

Secondary cuts are small additional cuts produced during the action of a stab cut. They are produced through a fold in the material during the main cut. If a knife draws fabric

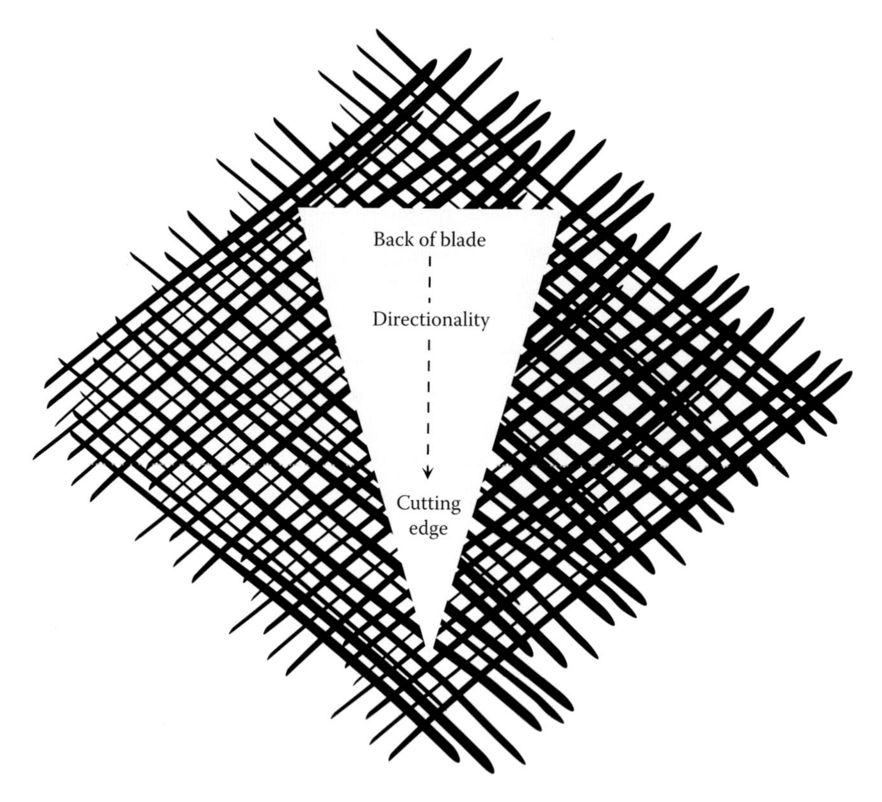

Figure 6.13 Stab-type cut.

into the wound, a fold will be produced. The fabric may then be cut at the fold, producing a small additional cut in line with the main severance (Figure 6.14). Another mechanism for secondary cuts is the stabbing implement penetrating folds of fabric in the course of the thrust.

Cuts through folds are a special type of damage that can occur in a stabbing scenario, especially in multiple stabs that arise in a frenzied attack. The process of stabbing may be dynamic, whereby the victim will move in response to avoid being stabbed and clothing may become bunched and folded during the entry and exit of the knife from the victim. This can result in cutting damage that is not straight, and when the fabric is folded flat, the length of the cut may not reflect the true width of the blade of the knife due to a fold being present in the area of cutting. Furthermore, when a knife penetrates folds in clothing, a single stabbing action can result in two or more cuts in the fabric. If items of clothing have moved during a stabbing action, this may increase the dimensions of the area of fabric damage. The damage observed needs to be considered in conjunction with the examining pathologist or medical examiner.

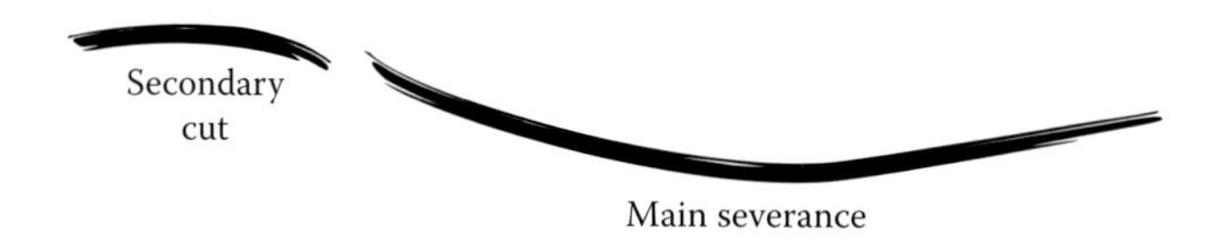

Figure 6.14 Secondary cut.

6.9 Simulations

Simulation experiments may be required in order to include or exclude a particular weapon as the source of the damage. It may also be necessary to attempt to simulate a particular scenario that has been postulated. Experience or expertise does not mean that the examiner has encountered a particular damage scenario. The Chamberlain case (Morling, 1987) illustrates the dangers inherent in assumptions about the cause of clothing damage. At the Chamberlain's original trial, the prosecution convinced the jury that the damage to the baby's jumpsuit was the result of scissor cuts or a knife, and not a dingo as the parents claimed. It was asserted that dingoes could not produce cuts with their teeth; however, experiments with dingoes were not performed. The Royal Commission into the Chamberlains' convictions conducted experiments with dingoes damaging jumpsuits of the same material as the original; they showed that dingoes could indeed create damage that could not be differentiated from the original damage.

It is sometimes not possible to choose one scenario over another (Taupin, 1998b). Often this is because of the innumerable variables involved in a crime scenario that are unknown or cannot be replicated. Sometimes it is because of the inherent limitations in damage analysis. The following case (Taupin, 1998b) involved competing scenarios by the prosecution and the defense:

> The defendant was having an argument with his ex-girlfriend and picked up a kitchen knife. An incident occurred that caused the knife to become embedded in the woman's right shoulder. Ambulance officers attended and pulled the knife out, so the identity of the weapon was not at issue. The woman said she blacked out and could not remember what happened. The defendant claimed that he threw the knife at the ex-girlfriend's dog but it accidentally hit her. The prosecution alleged that he deliberately stabbed her; this went toward intent and the seriousness of the charge. The scenario of stabbing at close range was not simulated because it is frequently encountered with stab-type injuries. The other scenario was simulated by purchasing a knife similar to the weapon (the blade of which was found to be bent on removal). A pork leg, providing a support medium similar to human flesh, was strapped to the chest area of a mannequin over a T-shirt. When the purchased knife was thrown at the mannequin from a distance, by either the handle or the blade, the knife penetrated the pork flesh, the T-shirt, and/or the body of the mannequin, and remained embedded. Features of the damage produced in the T-shirt were similar to stab-type cuts.
>
> The scenario proposed by the defense in this case was initially thought unlikely by knife-throwing experts consulted, such as circus managers. "Throwing" knives, such as those used in circus acts, are weighted in the blade for ease of penetration when thrown. Because the evidence knife was not weighted, it was believed that it would be unable to penetrate flesh when thrown. However, simulation experiments showed the knife in fact penetrated the hard plastic of the mannequin or up to 8 cm in depth of the pork flesh. The accused pleaded guilty to recklessly causing serious injury. The prosecution accepted this plea due to the difficulty of proving he "intentionally" caused serious injury.

Sometimes only one scenario is being proposed; if the scenario is not supported by the simulations, then the examiner may need to list impossibilities rather than likely scenarios.

Simulation damage may have to be performed on the damaged garment itself if it is considered that aging of the garment is important and the wear and tear cannot be reproduced. Depending on the situation, the examiner may elect not to perform simulations on the evidence garment, especially if the item is small, such as ladies' or men's underpants. A substitute garment or fabric should then be chosen to replicate the case garment as closely as possible.

Depending on the scenario, the garment may need to be held under a certain tension or have "backing." A roll of pork, using greaseproof paper to protect the garment, may be used to simulate flesh if the wounds are close to a fleshy area (Monahan and Harding, 1990). This is especially useful in cases involving stabbing or slashing to the body.

If a "suspect" weapon is provided, then simulation stabs or slashes should be performed on the test or simulated garment and the resulting characteristics compared with the "crime" damage. If a weapon is not provided but a class of weapon is indicated (e.g., a screwdriver is responsible for the damage), then controlled tests may be performed with an instrument representative of the class of weapon. Most weapons are mass produced, so the resultant damage may only be indicative of the class of weapon. It is generally not possible to attribute evidence damage to a unique source weapon or to give statistical probabilities about the likelihood of seeing the damage if it was caused by that weapon.

In all cases, the examiner should remember that variables such as body weight and strength of the individuals involved, their movement, the angle and type of thrust, and the position of the clothing in the crime event are unknown or cannot be replicated. Consequently, these limitations should be considered in any conclusion.

6.10 Physical Fit

The traditional notion of physical fit involves fitting together pieces of solid material in a jigsaw arrangement to complete or partially complete an object. However, we can extend this idea to clothing damage if it is suspected that there are pieces of the one garment for examination. A balaclava cut from the sleeve of a fleecy sweater is an example. This is useful evidence if the balaclava is dropped at the scene of an armed robbery and the sweater is found at the home of the suspect (in addition to any DNA evidence).

The finding of buttons at the crime scene and missing buttons on a suspect's shirt may also be valuable evidence. If buttons are torn off in a struggle, sometimes either garment fabric and/or button stitching yarns remain attached to them. It is thus possible to compare the buttons not only to the remaining buttons on the shirt but also to the shirt fabric and button stitching (see Figure 6.15). Similarly, other types of fastenings or laces and trims may be torn off at the scene and link the victim or the suspect to the scene. It is generally not possible to physically fit torn edges of a garment or thread in the manner of fitting cut or neat edges due to the irregular rupture of the tearing process in materials. Nevertheless, one can still associate torn fabric and yarns to their originating garment with pattern matching, fabric composition, and structure. It is feasible to fit together pieces of a garment and determine that no significant quantity of fabric is missing by the matching of rib lines and severance faces.

6.11 Glass Cuts

Glass cuts to clothing can often be found in cases of burglary, motor vehicle offenses, and other crimes where glass has been broken during the course of the crime. Glass cuts may include irregular cuts and punctures, with features of tearing, stab-type cuts, and slash-type cuts with discontinuities (Taupin, 1998b). A broken wine bottle may even

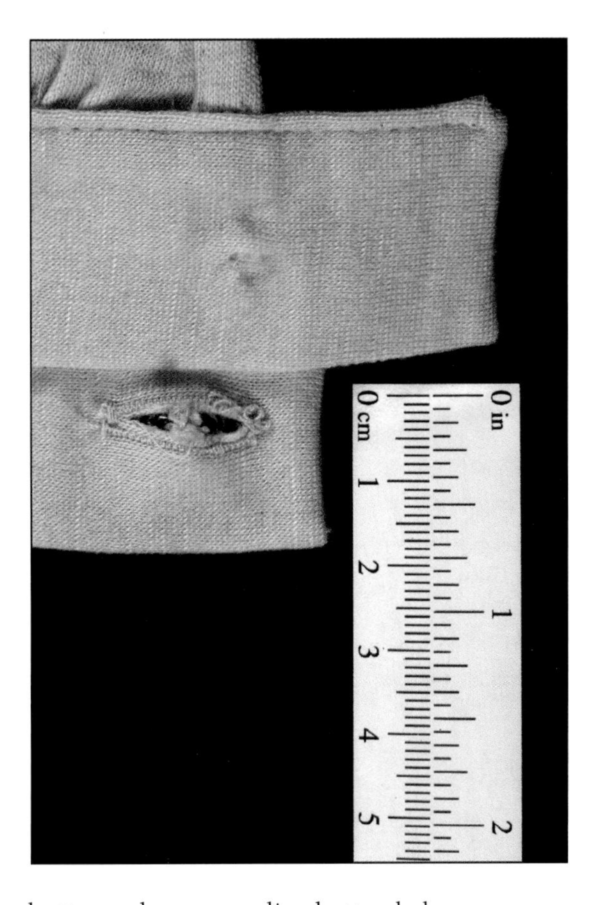

Figure 6.15 Missing button and corresponding button hole.

produce features in clothing that cannot be differentiated from stab cuts with a knife (Taupin, 1998b).

Freshly broken glass may have razor-sharp thin blades capable of cutting yarns and fibers neatly. However, broken glass also has broad surfaces that can catch fabric and result in tearing. Consequently, it is not unusual to see both cutting and tearing features in glass damage to clothing. Small pieces of fabric may also be completely cut away as a result of the shard or target moving while the cut is being formed.

Glass cuts may also have characteristics indicative of multiple shallow, sharp blades. The multiple cutting appears to occur due to the presence of multiple cutting surfaces, both on single shards of glass and on multiple shards of glass (Griffin, 2007).

6.12 Microbial Damage

Damage to clothing through microbiological activity is often found in burials. Fungal and bacterial activity play an important role in decomposing organic materials such as wool and cotton in clothing. Cotton, which is highly susceptible to microbial damage, is destroyed very quickly in soil burial situations. However, synthetic fabrics are less prone to microbial action. A mix of fibers in a garment may lead to preferential damage. "Pseudo" cuts may be produced in buried fabric. Dye-coated natural fibers may be protected, whereas undyed fibers may be subject to attack.

A study of woolen soldiers' uniforms buried in soil found that the length of time and relative access of microflora to the fibers had an important influence on the degradation process (Was-Gubala and Salerno-Kochan, 2000). The final decay of the uniforms took place after 5 to 6 weeks of burial in standardized bioactive soil.

6.13 Thermal (Fire and Heat) Damage

Thermally changed fibers, yarns, and fabrics composing clothing may be found in cases of arson, explosions, and road accidents. Fires are sometimes set to mask other crimes. Heat damage to materials may also be a component in establishing temperatures at different areas of a fire scene. Challenges in these cases include determining the original fiber types and establishing the thermal conditions during which the fibers undergo any change. Was the item briefly exposed to high heat and then baked at lower heat? Was the entire exposure to lower heat only? Was the heat exposure of long duration or only an instant?

Sometimes the garment has been almost totally consumed in the fire or the explosion; occasionally the garment has just been slightly burnt or partially consumed. Using techniques such as FTIR microscopy, the identification of, as well as the differentiation between, thermally changed fibers can be performed (Was, 1997). It is even possible to identify melted, decomposed, burned, and incinerated fibers (Was, 1997). Analysis using polarized light microscopy can provide complementary information about the original fiber type and about thermal changes to the fibers.

It is necessary to understand the nature of the garment's composition in heat and fire cases. The process of single fiber thermal degradation is different for various kinds of fibers (Was-Gubala and Krauss, 2006). Nylon, polyester, polyolefins, and several other synthetic fibers change primarily in terms of their physical state as the temperature increases (softening, contracting, and melting), although some undergo phase changes observable under polarized light microscopy. Chemical degradation (decomposition and burning) usually occurs after the melting point of the fiber is exceeded. Vegetable fibers change primarily in terms of their chemical structure when the temperature is increased, retaining their original shape until loss of mass results in disintegration. Wool and hair both melt and decompose, swelling across the fibers and bubbling with released gases, producing a distinctive smell and appearance. The sequence and initial appearance of these changes depends on the rate of heating and the initial temperature.

Garments initially change to a yellowish color then to a brownish color upon exposure to an open flame. Loosening of the structure of the fabric may be observed mainly in cellulosic material. Wool and most synthetic fibers form a hard crust (conflagration of fibers) on the surface. Direct exposure to an open flame will result in fibers varying in morphological appearance, compared with fibers exposed to a shock wave accompanying a vapor cloud explosion (Was-Gubala and Krauss, 2006).

A vapor cloud explosion occurs when a mixture of explosive gases becomes ignited and reaches a temperature between 800 and 1000°C. A heat wave from an explosion passes through a garment quickly, and consequently many fibers in the yarn comprising the garment remain unaffected. The relatively few affected single natural fibers (cotton and wool) will become burnt, showing brown to black ends (Was-Gubala and Krauss, 2004). Synthetic fibers melt and form conglomerates, and their fiber ends form the shape of a ball, bulb, or shovel (Was-Gubala and Krauss, 2004).

6.14 Firearm Damage

Shooting is a major cause of homicide in the United States, particularly for young people (Centers for Disease Control, 2004). Firearms are also the weapon of choice for suicide in this country. The number of firearm injuries is high compared to countries such as England and Australia, where the possession of firearms is restricted. The proportion of firearm damage to clothing will also be greater in the United States compared to stabbing damage, as discussed earlier.

Firearms commonly encountered in crimes discharge either bullets or pellets. Pistols, rifles, and submachine guns discharge bullets and shotguns discharge pellets. Very little exists in the literature on either bullet or pellet damage to clothing. Comparatively more information exists on firearm wounds to the body (DiMaio, 1998; Froede et al., 1982).

Shotgun wounds differ from those of other missiles because the spectrum of wound severity is large, owing to the fact that the pellets scatter as they travel. Close-range shotgun wounds can be as destructive as those from a high-velocity rifle; but at longer range distances, there may be only minimal injury. Shotgun pellet patterns increase in size as the muzzle-to-target distance increases. The length of the shotgun also affects the pellet pattern (Speak et al., 1985).

Similarly, damage to clothing may reflect the distance of the shotgun from the clothing. Figure 6.16 shows a close-range shot from a sawed-off shotgun to a padded jacket. Figure 6.17 shows a spread of pellet from the same shotgun, indicating that the fire was from a further distance than the close-range fire. This assumes, however, that there has been no intervening medium.

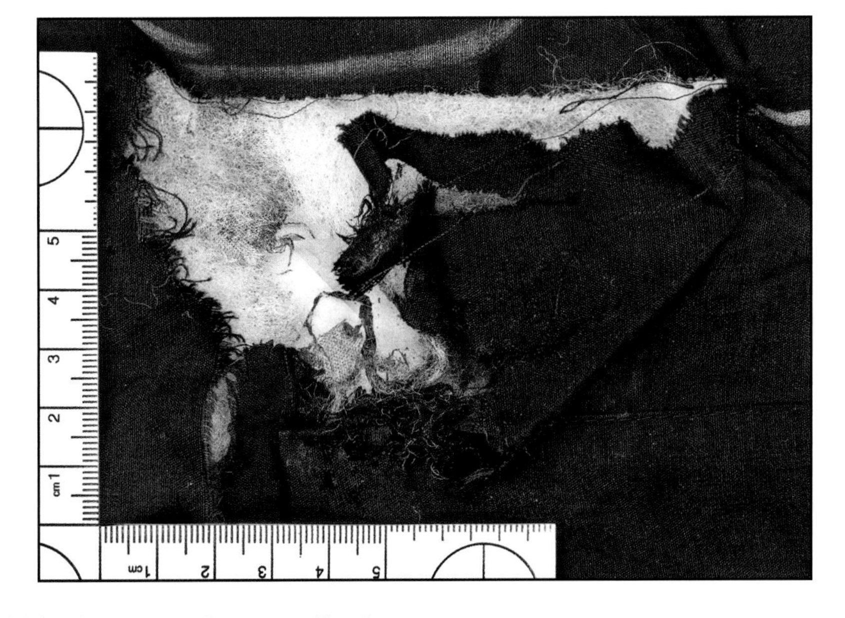

Figure 6.16 Close-range shotgun pellet damage.

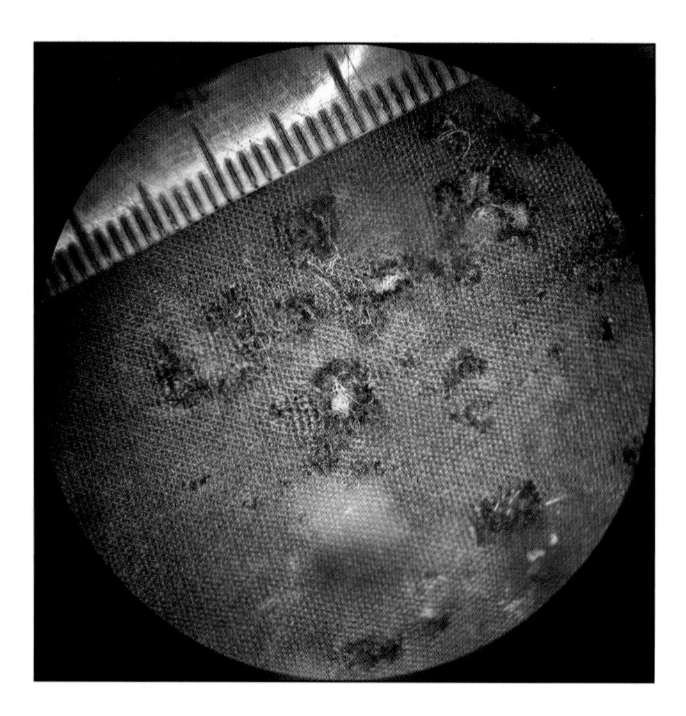

Figure 6.17 Medium-range shotgun pellet damage.

The composition of the garment determines the appearance of any firearm damage. Contact bullet wounds through clothing may cause tearing and/or melting of the fabric. Medium- and large-caliber weapons can cause tears with a stellate or cruciform (cross-like) appearance for cotton or cotton-mix woven fabric, where the tear travels along both the warp and the weft directions. A complete stellate defect has three or more tears radiating away from a central hole; a partial stellate defect has only two tears radiating away in opposite directions from the central defect (Alakija et al., 1998; see Figure 6.18).

Stellate tears are conventionally believed to result from contact or near-contact firearm entrance shots. One of the few papers in the literature regarding experimental firearm damage to clothing investigated firearm damage to cotton weaves and knits (Alakija et al., 1998). They found that a .30-30 Winchester lever-action rifle produced stellate tears in the clothing on contact with the firearm muzzle, and at distances from the muzzle up to 8 cm. The .22 pistol produced stellate tears on contact; non-stellate defects were produced at distances of 2 cm or greater. The .22 bolt-action rifle did not produce stellate tearing at any distance.

"Burn holes" may be produced in synthetic fabrics when the firearm muzzle contacts the garment (DiMaio, 1998). The heat of the gases from the weapon's muzzle flash may cause synthetic material to melt, producing large circular holes, usually with scalloped margins, or it may simply cause melting of some of the fibers at the defect margins.

When a bullet or bullet fragment exits the body, it may not have enough energy to completely penetrate all the layers of fabric and may get stuck in the fabric. It may also strike the fabric without penetrating, then slip out or remain inside the clothing. If a shirt sleeve, but not the jacket sleeve, shows evidence of a bullet exit, the examiner should look at the jacket sleeve closely for small fabric defects such as deformation of fibers or an area of distended weave that may indicate bullet impact. Traces of bone, tissue, or fibers from

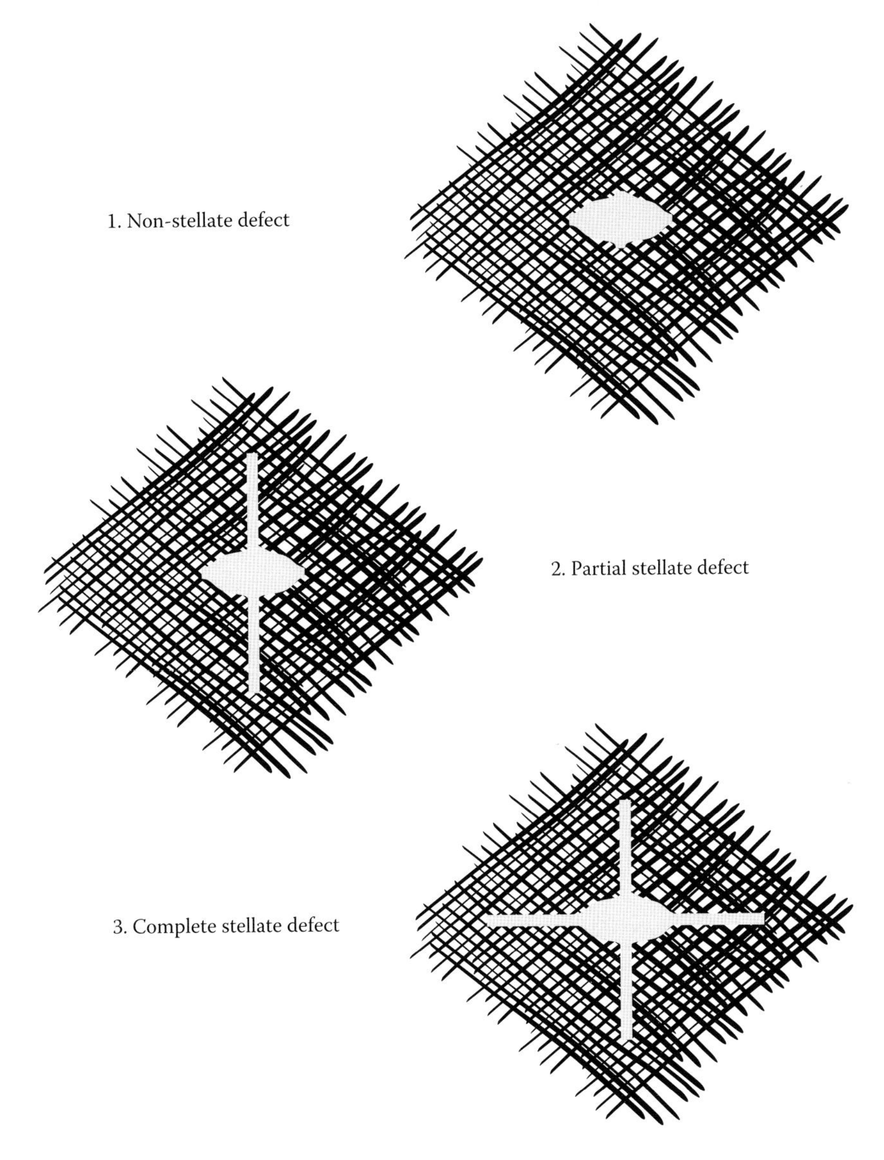

Figure 6.18 Stellate tear defects.

the preceding layer may have been deposited upon nonpenetrating impact. All potential bullet holes should be examined for traces of fabrics from other layers; this can provide information about the bullet path.

In shooting cases, it is important to determine which fabric defects are from a bullet entrance and which are from a bullet exit, because this can assist with reconstruction of events. It may also be important to determine whether a bullet entered the fabric after passing through another target or after striking another object and then ricocheting. A bullet may first enter an arm, then exit and enter the chest. This produces a primary entrance defect, an exit defect, a secondary entrance defect, and a possible second exit. In correlating the defects in the clothing with gunshot wounds in the body, it is common to find that a bullet with a secondary entrance does not penetrate as deeply in the body.

Not every bullet that strikes a garment also strikes the body of the person wearing it. A bullet may graze the surface or pass through loose folds or flaps. It may exit through another part of the fabric or, in the case of a flap such as an open panel of a shirt or jacket, it may simply fall to the ground. A bullet may also strike a hard object on a clothing item and ricochet. A bullet entrance with no corresponding wound or a separate bullet exit with no surrounding blood or tissue may indicate a shot that did not enter the body. If blood is observed, the examiner should evaluate the bloodstains in the surrounding area to determine whether they emanate from the bullet defect of interest or from another source of bleeding.

6.15 Other Textiles

It has been shown that paper towels behave similarly to clothing fabric in response to the damaging action of cutting tools (Causin et al., 2005). However, because paper towels have a less regular structure compared to most clothing materials, there are more limitations to any conclusion. Cuts and tears may be differentiated, and it may be possible to infer the class of tool that caused the damage, but directionality cannot be determined (Causin et al., 2005). Paper towels can be considered as a type of felt, where the fibers are compressed together to form the textile. Other non-woven fabrics would be expected to yield similar results.

6.16 Limitations

It is essential that all relevant background information regarding the case is ascertained. This may be especially important if the garment incurred additional damage subsequent to the crime event. This type of damage typically occurs when hospital personnel or others must remove the victim's clothes (e.g., cutting off clothes with scissors) or if the distraught victim tears the clothes. Further damage may occur when the garment is removed at autopsy, or in packaging and transport of the garment to the laboratory, especially if the garment has damage that may be readily extended. Such post-event damage is typically recorded in crime scene and autopsy photographs, which should be examined in conjunction with the clothing.

6.17 Glossary of Terms

Cut: A severance caused by a sharp implement
Directionality: Profile of cut indicates the orientation of cutting edge or back of blade
Hole: Defect in material that has penetrated all yarns of material in the area; often produced by firearms
Hospital-type damage: Created by medical personnel in order to examine a patient; most often caused by scissors and tearing
Knit: Continuous thread formed by interlocking loops
Nick: Small cut or notch, sometimes at an end of a cut
Pilling: Small balls of fibers formed by abrasion through normal wear and tear, often attached to garments
Planar array: A cut where ends of fibers or yarns line up in the same plane

Selvedge: The outer finished edge of a fabric
Simulation: An experiment designed to reconstruct a proposed scenario as accurately as possible
Snippets: Short segments of yarn (between 0.5 and 2.0 mm) that have been cut completely from the fabric; created if a knit fabric is cut at an angle to the thread direction
Stoppages: Indicates the areas where scissor blades have closed and opened to commence a new cut
Tear: A severance caused by pulling with some force
Tongues: Tears or scissor cuts may give rise to "tongue-like" protuberances
Tufts: Collection of snippets or loops of yarn from a knit
Weave: Formed by interlacing two sets of yarns at right angles

6.18 References

Alakija, P., Dowling, G.P., and Gunn, B., Stellate clothing defects with different firearms, projectiles, ranges and fabrics, *J. Forensic Sci.*, 43(6), 1148–1152, 1998.
Boland, C.A., McDermott, S.A., and Ryan, J., Clothing damage analysis in alleged sexual assaults — The need for a systematic approach, *Forensic Sci. Int.*, 167, 110–115, 2007.
Causin, V., Marega, C., and Schaiavone, S., Cuts and tears on a paper towel: A case report on an unusual examination of damage, *Forensic Sci. Int.*, 148, 157–162, 2005.
Centers for Disease Control, *National Vital Statistics Reports*, 52(21), June 2004.
Costello, P.A. and Lawton, M.E., Do stab-cuts reflect the weapon which made them? *J. Forensic Sci. Soc.*, 30, 89–95, 1990.
DiMaio, V., *Gunshot Wounds: Practical Aspects of Firearms, Ballistics and Forensic Techniques*, 2nd ed., CRC Press, Boca Raton, FL, 1998.
Froede, R.C., Pitt, M.J., and Bridgemon, R.R., Shotgun diagnosis: "It ought to be something else," *J. Forensic Sci.*, 27(2), 428–432, 1982.
Green, M.A., Stab wound dynamics — A recording technique for use in medico-legal investigations, *J. Forensic Sci. Soc.*, 18, 161–163, 1978.
Griffin, H., Glass cuts, in *Forensic Analysis on the Cutting Edge: New Methods for Trace Evidence Analysis*, Blackledge, R.D., Ed., Wiley-Interscience, Hoboken, NJ, 2007.
Hearle, J.W.S., Lomas, B., and Cooke, W.D., *Atlas of Fiber Fracture and Damage to Textiles*, 2nd ed., CRC Press, Woodhead Publishing, Cambridge, England, 1998.
Hunt, A.C. and Cowling, R.J., Murder by stabbing, *Forensic Sci. Int.*, 52, 107–112, 1991.
Johnson, N., Physical damage to textiles, in *APPTEC Conference Proceedings*, Canberra, Australia, 1991.
Knight, B., The dynamics of stab wounds, *Forensic Sci.*, 6, 249–255, 1975.
Maxwell, R., Trotter, C., Verne, J., et al., Trends in admissions to hospital involving an assault using a knife or other sharp instrument, England, 1997–2005, *J. Public Health*, 29(2), 186–190, 2007.
Monahan, D.L. and Harding H.W., Damage to clothing — Cuts and tears, *J. Forensic Sci.*, 35, 901–912, 1990.
Morling, T.R., *Royal Commission of Inquiry into Chamberlain Convictions*, Government Printer of the Northern Territory, Darwin, Australia, 1987.
Ormstad, K., Karlsson, T., Enkler, L., et al., Patterns in sharp force fatalities — A comprehensive forensic medical study, *J. Forensic Sci.*, 31(2), 529–542, 1986.
Pelton, W., Distinguishing the cause of textile fiber damage using the scanning electron microscope (SEM), *J. Forensic Sci.*, 40 (5), 874–882, 1995.
Rouse, D.A., Patterns of stab wounds: A six year study, *Med. Sci. Law.*, 34, 67–71, 1994.
Speak, R.D., Kerr, F.C., and Rowe, W.F., Effects of range, calibre, barrel length, and rifling on pellet patterns produced by shotshell ammunition, *J. Forensic Sci.*, 30(2), 412–419, 1985.

Stowell, L. and Card, K., Use of scanning electron microscopy (SEM) to identify cuts and tears in a nylon fabric, *J. Forensic Sci.*, 35, 947–950, 1990.

Taupin, J.M., Damage identification — A method for its analysis and application in cases of violent crime, *Proceedings of the 14th Meeting of the IAFS*, Tokyo, 1996.

Taupin, J.M., Arrow damage to textiles — Analysis of clothing and bedding in two cases of crossbow deaths, *J. Forensic Sci.*, 43(1), 205–207, 1998a.

Taupin, J.M., Testing conflicting scenarios — A role for simulation experiments in damage analysis of clothing, *J. Forensic Sci.*, 43(4), 891–896, 1998b.

Taupin, J.M., Comparing the alleged weapon with damage to clothing — The value of multiple layers and fabrics, *J. Forensic Sci.*, 44(1), 205–207, 1999.

Taupin, J.M., Clothing damage analysis and the phenomenon of the false sexual assault, *J. Forensic Sci.*, 45(3), 568–572, 2000.

Taupin, J.M., Adolf, F.P., and Robertson J., Examination of damage to textiles, in *Forensic Examination of Fibers*, 2nd ed., Robertson, J. and Grieve, M., Eds., Taylor and Francis, London, 1999.

Upkabi, P. and Pelton, W., Using the scanning electron microscope to identify the cause of fiber damage. Part 1: A review of related literature, *Can. Soc. Forensic Sci. J.*, 28, 181–187, 1995.

Was, J., Identification of thermally changed fibers, *Forensic Sci. Int.*, 85, 51–63, 1997.

Was-Gubala, J. and Krauss, W., Damage caused to fibers by vapour cloud explosions, *Forensic Sci. Int.*, 141, 77–83, 2004.

Was-Gubala, J. and Krauss, W., Damage caused to fibers by the action of two types of heat, *Forensic Sci. Int.*, 159, 119–126, 2006.

Was-Gubala, J. and Salerno-Kochan, R., The biodegradation of the fabric of soldiers' uniforms, *Sci. Justice*, 40, 15–20, 2000.

7 Human Biological Evidence

Physiological fluids and biological material are the most common types of physical evidence found in violent crime. The advent of DNA typing, with the possibility of individualization, has increased their importance. This chapter will discuss the examination of blood, semen, and other human body substances that are deposited on clothing during the commission of a crime. Patterns of these substances (usually blood) are discussed in Chapter 5. Their composition, especially if they can be DNA profiled, may be crucial in including or excluding any suspect and is discussed in this chapter. Although hairs are of a biological origin and may be DNA typed, they behave like fibers and are thus discussed in Chapter 8.

The main objectives of biological evidence analysis are identification (or classification), individualization (DNA typing), and reconstruction. Classification of the evidence is determining whether it is blood, semen, or another bodily substance. Individualization of biological evidence is based on DNA typing, which is highly discriminating and has the power to attribute biological evidence to an individual with an extremely high degree of probability. Reconstruction of biological evidence is the interpretation of the pattern of the evidence, such as blood pattern analysis or the location of the semen stain with reference to the proposed crime scenario.

Presumptive tests are used to screen stains or deposits on items of clothing that the examiner suspects may be the substance of interest; for example, blood or semen. These tests are sensitive but usually not specific for that material. Identification will require confirmatory tests. Confirmatory tests are specific but are more time consuming and use more sample. Presumptive results should determine the most appropriate action for further testing and conservation of sample.

It is not always necessary to confirm the type of body fluid on the clothing item; in fact, doing so may be detrimental to subsequent DNA analysis if presumptive or confirmatory tests consume or dilute the fluid's cellular constituents. It may be more important in the investigation to identify the donor of the biological sample. Stamps and envelopes are examples of items for which the body fluid type (saliva) can be inferred from the adhesive areas on the item.

Sampling will be an issue whenever there is multiple blood or other biological stains on a garment. How many stains should be sampled and how many tests should be performed? As with any case, the decision to sample or perform a particular test should be based on background information regarding the alleged crime and the scientific method using hypothesis testing. The number of tests should be minimized to ensure that sufficient sample remains for DNA analysis and any retesting that may be required.

Determining body fluid origin (body location or individual) is one of the most difficult tasks for the forensic examiner, especially if there is a potential mixture. Interpretation may be ambiguous, especially in mixtures of DNA from two or more individuals, degraded samples, or a limited sample. The examiner should be informed as much as possible, but the examiner must often rely on the information suggested by the stains themselves and how they were deposited. In the case of degraded stains or limited sample, which is nevertheless potentially significant, the realities of the evidence may dictate that these samples be collected and examined, even if the results are not as specific as desired. The examiner should convey any problems or ambiguities to the investigators and to the court.

7.1 Blood

Blood is usually the easiest biological material to locate on clothing due to its reddish-brown color. It may dry to a crusty stain or a light smear depending on the quantity. Blood may soak through the fibers of the material and stain the opposite surface of the garment (such as the inside lining of a jacket). It may also soak through the various garments of the wearer. A visible stain can be readily tested with a presumptive test. If the stain is too light to see with the naked eye, a stereomicroscopic examination may be performed. Particles or flakes of blood may be visualized in this way. If the garment is composed of a dark material that could disguise bloodstains (such as a dark, reddish-brown T-shirt), then enhanced lighting or a variable light source may be needed (see Chapter 3).

Figure 7.1 shows a garment on which the search for blood should be relatively quick and easy. It depicts a long-sleeved plain white woven shirt, clean and unstained except for reddish-brown spots (confirmed to be blood after testing), circled with a yellow marker. Figure 7.2 shows a closer view of one of the blood spots. The color of the spot is obvious due to the white background of the shirt; it can be readily measured and its shape determined; and how it lies on the weave may be discerned under the microscope. Some items of clothing present greater difficulty to the clothing examiner; for example, dark-colored garments, patterned, stained with other fluids, damaged, and composed of folds or numerous layers.

7.1.1 Testing for Blood

Most *presumptive* tests for blood are color change tests that are based on the peroxidase-like properties of hemoglobin present in red blood cells (Gaensslen, 1983). Hemoglobin

Figure 7.1 Photograph of a long-sleeved white shirt with blood spots circled in yellow marker.

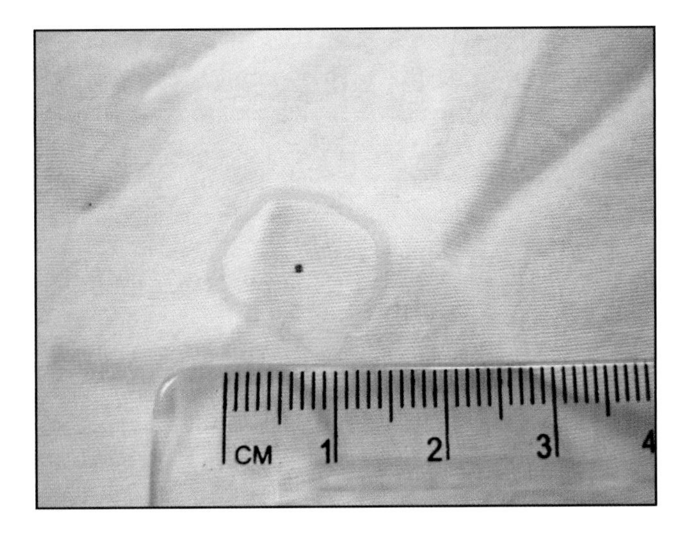

Figure 7.2 Close-up photograph of a blood spot on the shirt from Figure 7.1.

will catalyze the oxidation by peroxide of a number of organic compounds to yield colored products. Hydrogen peroxide is most commonly used as the peroxide. These tests can be done directly on the garment or by a transfer technique. A transfer technique is preferred, unless testing extremely small samples or large areas. The Hemastix™ Test (using tetramethylbenzidine, TMB) is one of the simplest and most sensitive and can be used at scenes as well as on clothing. It consists of a plastic strip with an orange-colored reagent-treated small pad at the end that is rubbed lightly on the suspect bloodstain. The pad is then moistened with distilled water. A color change to dark green will occur in the presence of blood. The Kastle-Meyer (KM) test is a two-step test with the reagent phenolphthalein, an acid–base indicator, added to filter paper that has been rubbed on the sample. On addition of an oxidizing agent (hydrogen peroxide in a 3% solution), the transferred sample on the filter paper will turn pink in the presence of blood. Similarly the leucomalachite green (LMG) test produces a green color in the presence of hemoglobin (see James and Nordby, 2005, for presumptive blood tests).

The luminol reagent produces light by undergoing a chemical reaction when it is oxidized in alkaline solution, a process called chemiluminescence (Gaensslen, 1983). Hemoglobin acts as an "accelerator" in these reactions. The luminol test is a presumptive blood test sensitive to 1 in 100,000 dilution (Tobe et al., 2007). It is used predominantly at crime scenes to elucidate bloodstains or blood patterns that are not visible to the naked eye due to dilution, such as in attempts to clean up the scene. The examiner should photograph the luminescence to record the pattern or result. Copper and some other metal salts, as well as plant peroxidases, will give false-positive reactions. It is possible to obtain DNA profiles from latent bloodstains after using the luminol reagent (Manna and Montpetit, 2000). The clothing examiner must observe the item of clothing in darkness, such as an enclosed room in a laboratory with the lights off, because the reaction is light producing.

Bloodstains that have been diluted by machine washing may not be visible, especially on dark-colored clothing, and thus a chemical reagent such as luminol may be the only method for developing latent bloodstains. Luminol can be used to locate latent blood transfer patterns and other larger bloodstain patterns that may have been washed or cleaned up. However, the examiner should use caution when interpreting such diffused or diluted bloodstain patterns

(Adair and Shaw, 2005). Luminol should also not be used to interpret the small blood spatters often created as a result of beating, stabbing, or shooting events (Pex, 2005). The examiner should avoid spraying a chemical directly onto a garment, such as is employed in luminol enhancement, if other techniques can be used. The process alters the remaining stains and deposits, with attendant loss of information, and renders any reexamination difficult.

The ideal presumptive test for blood is one that is specific to blood as much as possible, has a high sensitivity, and will not damage underlying DNA on the garment. Hemastix™, KM reagent, and LMG have been reported to be detected up to a 1 in 10,000 dilution, show false-positives to only a small number of substances other than blood, and achieve DNA amplification even when directly tested on the sample.

Confirmatory tests for blood are crystal tests and immunological tests using antibodies. Crystal tests, such as the Takayama test, are simple but may use more of the suspect material than is desired. A sample of the stained area of the garment is placed on a microscope slide and covered by a cover slip; a drop of the reagent (such as pyridine hemachromogen) is added under the cover slip. The sample is then examined using microscopic magnification. If distinctive crystals are produced, then the sample is positive for blood (Gaensslen, 1983).

Immunological confirmatory tests use similar techniques to species identification. Antibodies for various animal bloods are commercially manufactured and are used as a testing chemical for a specific animal antigen. The Ouchterlony double immunodiffusion method (Gaensslen, 1983) involves placing extracts of the bloodstain to be analyzed with specific antisera. If the bloodstain contains antigens corresponding to the specificity of the antisera, the antibodies bind to their antigens and the complex precipitates in a visible line on the gel. Using antibodies specific for human hemoglobin combines the confirmatory test for blood with the human species test into a single procedure.

Recently, new test kits have been marketed as confirmatory tests for human blood (Reynolds, 2004). These kits are especially useful at crime scenes due to their portability but may also be useful in the laboratory for clothing examination. They use an immunochromatographic technique that reacts with human hemoglobin.

Many laboratories today perform (1) a presumptive test for blood and then (2) DNA typing on the suspect sample. It is inferred that the sample is of human origin if a profile is obtained, because DNA typing is higher primate specific. A positive presumptive test for blood infers that the sample is human blood. It must be remembered that this is an indirect test; if it is imperative that a sample be confirmed to be blood, then a confirmatory test should be performed. If the size of the sample is an issue, then a sample of the extract used for DNA typing may be used for an Ouchterlony test for human blood.

Although the search for blood on clothing is often straightforward due to the amount of blood shed, in some cases crucial bloodstains have been overlooked due to the small quantity and/or difficulty in discerning the blood on dark clothing. The following case from England illustrates the problems that may face a clothing examiner in a high-profile crime (Rawley and Caddy, 2007):

> A 10-year-old boy, Damilola Taylor, died in 2000 on a London housing estate as a result of a stab wound to his thigh. The wound was possibly caused by a broken beer bottle and there was extensive blood loss at the scene. During the first trial of four boys in 2002, no forensic evidence existed. Two of the boys were found not guilty, and the charges were dropped against the other two. During a second police investigation, clothing belonging to two brothers, Danny and Rickie Preddie, were submitted for reexamination at a different forensic laboratory from that

which had examined all the original clothing items (more than 400 articles). A white training shoe belonging to Danny Preddie contained a small drop of blood, with directionality from above. This drop was DNA profiled and found to match the profile of Damilola Taylor. The blood drop had not been discovered in the first examination using the naked eye and KM presumptive tests. However, photographs of the shoe from the initial examination showed a "drop" on the shoe. A bloodstain was also found within the ribbing of a cuff of a sleeve of a black windjacket belonging to Rickie Preddie. Again, this stain matched the DNA profile of Damilola Taylor. On the first examination, no blood was detected on the windjacket. A general KM screen had been performed on it, with negative results. On reexamination, reddish-brown particles could be seen under magnification of 40 times on the cuff; these particles were found to be KM positive. The discovery of the two bloodstains led to prosecution of the two brothers, and they were eventually found guilty of manslaughter in 2006.

A government review (Rawley and Caddy, 2007) cited human failure, rather than systemic failure, as the reason the two relevant bloodstains were not found in the first examination. The reviewers identified a conflict between the pursuit of excellence and the demand for urgent results. No doubt the large number of clothing exhibits, nearly all marked "urgent," and the high profile of the case contributed to the pressure of the first examination. This apparent quandary will not surprise the many forensic scientists who are called upon to do work of high quality while under pressure to produce results promptly. A solution could be an impact-based priority system that allows the scientist to focus on the most useful work at the outset but not do everything at once. All clothing examination should be performed using the scientific method and hypothesis testing — that is, formulation of a hypothesis and performing a scientific experiment using appropriate methods. Blood from a stab wound to a victim may result in transfer of very small quantities of blood to the clothing of an assailant even though there may be extensive blood loss at the scene; the victim may bleed to death some time after the initial assault, well after the offender has absconded.

7.2 Semen

The presence of semen confirms sexual activity of a specific male; however, it does not determine whether such activity was by consent, nonconsent, or masturbation. The ejaculate volumes of human males range from 2 to 6 ml and contain a mean value of about 100 million sperm cells per milliliter (Jones, 2005).

The spermatozoa in the semen carry the individualizing traits of male DNA; thus, DNA profiling is generally used to type semen. The head of the sperm contains the cell nucleus, which is packed with DNA. The anterior portion of the head is capped with the acrosome, rich in enzymes to assist in penetrating the cell wall of the female egg during fertilization. A flagellated tail is attached to the head. The tail may be readily separated from the head and is not often seen on dried semen stains on clothing.

Dried semen on clothing may appear as a yellowish-white crust if it is undiluted or in sufficient quantity, so it may be readily visible. On white cotton garments, semen stains can have an off-white appearance. They tend to be stiff and turn yellow with age at room temperature.

The most common area for semen sampling in a rape case is the inside crotch of the underpants of the female victim, because this corresponds to vaginal drainage from internal ejaculate. Sometimes there is external ejaculation onto sweaters, pants, jackets, and other clothing items. The victim may also have wiped or spat ejaculate onto clothing, or sat on the edge of a nightgown or tail of a shirt, resulting in drainage.

The location of semen is not as simple as that of blood, because the color and consistency of dried semen can be confused with other body fluid stains. Untreated semen usually shows quite strong fluorescence, and ultraviolet light may locate the stains (see Chapter 3 for assisted light screening techniques), depending on the background luminescence of the clothing material and whether the clothing fabric quenches (absorbs) the fluorescence of the deposits. It should be noted that UV light can, at short wavelengths, degrade DNA, but long-wave UV light has little or no effect on subsequent STR (short tandem repeat) profiling; so examiners should be careful in their choice of light sources. For example, UV light at 254 nm is used to irradiate purchased unused plastic tubes to block DNA replication of any introduced DNA in manufacture. Fluorescence semen detection methods are attractive because they provide a rapid, nondestructive way of locating stains on large items and on a large number of items (Kobus et al., 2002). Because many types of materials, including foods, beverages, and oils, fluoresce, the presence of a fluorescent stain does not confirm semen; nor does a lack of fluorescence necessarily exclude the presence of semen, because some fabrics and fabric additives quench fluorescence (absorb the emitted light), so that none is observed.

7.2.1 Testing for Semen

Chemical presumptive tests can also be used, together with visual examination, to locate semen stains on clothing. The most commonly reported chemical presumptive test is based on the fact that the highest levels of the enzyme seminal acid phosphatase are found in semen. A color reaction occurs with the color developer brentamine fast blue (or fast black), which turns an intense purple within 2 minutes if semen is present (James and Nordby, 2005). Semen stains on clothing may maintain their seminal acid phosphatase activity for years if the clothing is kept dry and free from bacterial attack or other environmental insult. Seminal acid phosphatase is water soluble, so the enzyme can be transferred to filter paper easily.

If the semen stain, or a possible semen stain, is visible, a small piece of dampened filter paper can be pressed against it. When a drop of the brentamine fast blue reagent is added, a color reaction will indicate the presence of semen. Figure 7.3 depicts the purple color, showing positive results on dampened filter paper pressed against semen stains and sprayed with reagent, at 1 minute reaction time. Larger areas, even whole garments or bed sheets, can be searched using large sheets of dampened blotting paper pressed against the item and then spraying the reagent on the paper once the paper has been removed. The examiner should mark the paper to correspond with the area(s) of the item tested (see Figure 7.4). When a weak to moderate positive reaction is observed, the examiner must consider the presence of other body fluids, the time since the alleged sexual activity, and the fabric type of the garment.

A rapid presumptive seminal fluid test is the Phosphatesmo KM Paper® kit, which is a staining kit for acid phosphatases that behaves similarly to the screening acid phosphatase (AP) tests described above, in that the kit paper will turn purple in the presence of semen (Khaldi et al., 2004). Other test kits, such as a PSA kit (described below), have been investigated by these authors. The Polilight® (an alternate light source) is comparable in sensitivity with the AP chemical test (Vandenberg and van Oorschot, 2006).

Table 7.1 lists the common methods for locating semen on clothing.

Once the stain is determined to be presumptive positive for semen, the next step is to confirm the presence of spermatozoa. This will also confirm the presence of semen. A section of the stain is cut out and extracted in distilled water, and a drop of the extract placed

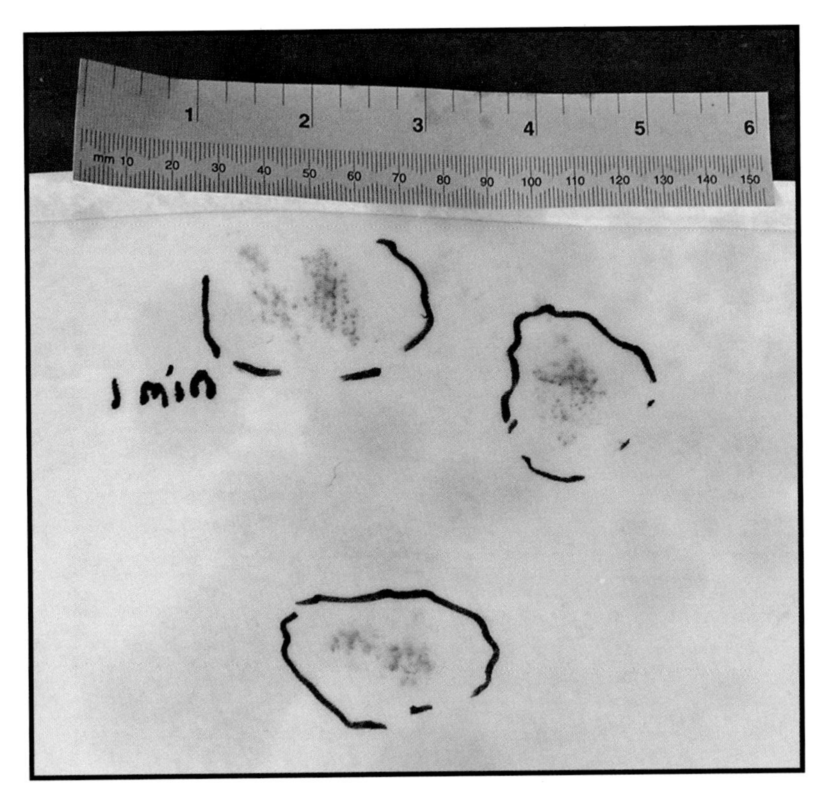

Figure 7.3 Photograph of semen presumptive positive areas on filter paper at 1 minute reaction time, denoted by purple staining.

on a slide and dried. This is then differentially stained to elicit the microscopic presence, shape, and form of any spermatozoa. Staining techniques include eosin/hematoxylin and Christmas tree (Shaler, 2002). The DNA-rich head of the spermatozoa is stained a blue-violet with hematoxylin while the cap and the tail is stained a pinkish-red with the eosin. The Christmas tree stain is popular in England and the United States and also uses differential staining. Human spermatozoa are distinct in size and shape; the examiner should be proficient in differentiating human from animal spermatozoa. Often the extract for the microscopic examination is also used for DNA profiling.

Extraction of the stain in water will usually involve mechanical mixing (vortexing) or sonication. Old or denatured seminal stains may require special or different techniques for removal of spermatozoa from the clothing material. Threads may need to be teased out from a cut stain when placed on a microscope slide and moistened with water. The fibers or yarns are then removed and the water is allowed to dry on the slide; differential staining then follows.

Sometimes it is extremely difficult and taxing to locate sperm on a slide swamped with nucleated epithelial cells from the victim. Extracts from clothing with a large number of epithelial cells may occur in vaginal–seminal mixtures, oral–seminal mixtures, and rectal–seminal mixtures, all typically found in the crotch area of the victim's underpants. These extracts may be treated with Proteinase K (Pro K) to selectively lyse the nucleated epithelial cells, leaving the sperm heads intact (Jones, 2005). Searching of slides for sperm is consequently easier and more efficient and the chance of missing a single or a few sperm

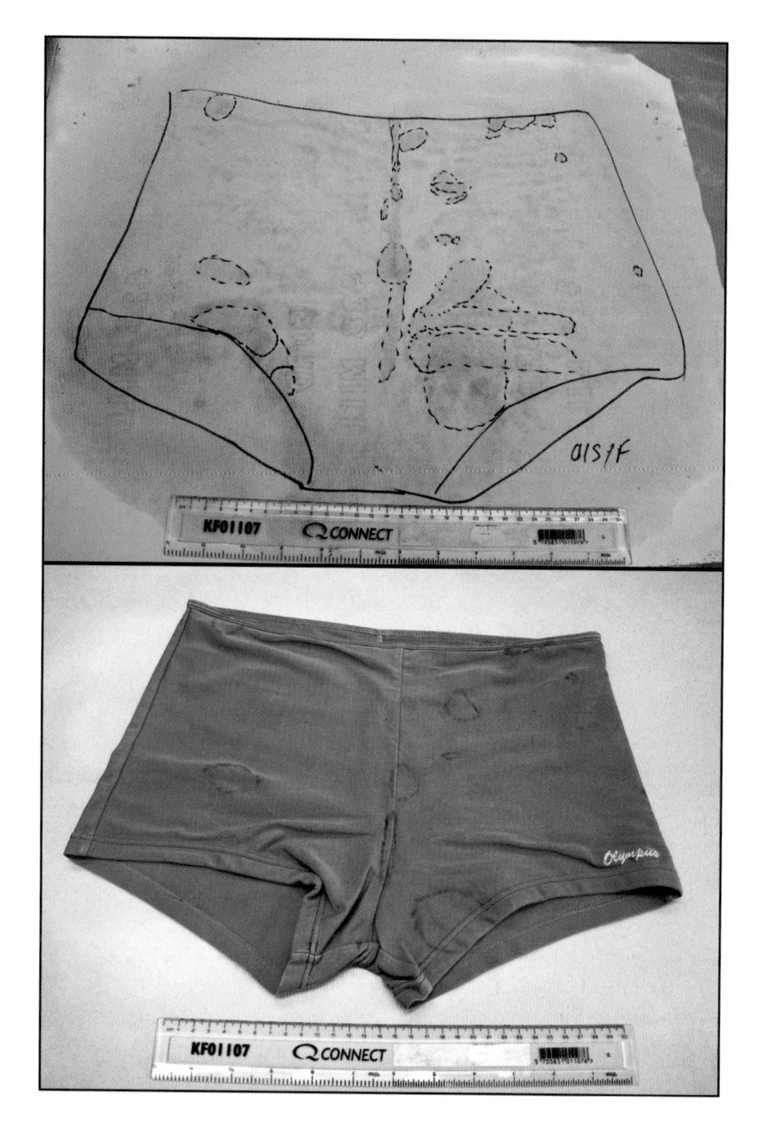

Figure 7.4 (Top) Photograph of blotting paper marked with the outline of a pair of ladies' underpants (boy-leg style). The paper has been pressed on the wetted underpants and sprayed with AP reagent. Positive AP areas (semen presumptive) are denoted by the purple staining. (Bottom) Photograph of ladies' underpants of above. Some semen presumptive positive areas have been marked on the garment by aligning with the markings on the blotting paper.

Table 7.1 Methods for Semen Location

Visual: Items are examined for stains visible to the naked eye that appear as a crusty white to faint yellow.
Physical: Areas of seminal stains on cloth may have a different texture from the rest of the material.
Chemical: Reagent-impregnated paper is used to map areas for the presence of acid phosphatase.
Fluorescence: Ultraviolet light is used to exploit the fluorescent properties of semen.

heads reduced. This procedure should be applied in cases where there is sufficient material for both biological fluid testing and DNA analysis.

Prostate-specific antigen (PSA or p30) is present in high amounts in seminal fluid, including that of males who have been vasectomized or have abnormal spermatozoa levels. Validated as a marker for the presence of seminal fluid (Hochmeister et al., 1999), PSA is a glycoprotein produced by the prostate gland. The PSA test is often performed to support the presence of azoospermic semen (semen that lacks spermatozoa), following a positive presumptive test for semen but an absence of spermatozoa in the extracts. A positive PSA test may also explain no DNA result. Test kits are available (James and Nordby, 2005); some are used in medical diagnostic tests as a marker for prostate cancer. These kits are immunochromatographic membrane tests. Prostate-specific antigen is known now not to be specific to the prostate only and perhaps should be renamed as "prostate-secreted antigen."

Semen stains may concentrate their components differentially according to the type of material on which they are deposited. Absorbent clothing fabrics like cotton tend to concentrate the water-soluble components, such as seminal acid phosphatase and p30, in the periphery of the stain, whereas the spermatozoa tend to stay in the middle. It may be necessary to sample a few sections of the seminal stain to obtain requisite spermatozoa for a complete DNA result.

It is now possible to perform DNA analysis on semen from vasectomized males even though no spermatozoa are present. Concentrated azoospermic semen samples have been analyzed using Y-STR technology and DNA obtained from the epithelial and/or white blood cells that were present (Shewale et al., 2003; Soares-Vieira et al., 2007). It has also been possible to separate spermatozoa from epithelial cells using laser microdissection technology (Sanders et al., 2006). This is useful when separation of the suspect's spermatozoa from the victim's own cells is incomplete.

Occasionally, clothing and bedding have been washed or laundered after the sexual assault, before police have had a chance to obtain them for forensic examination. Although machine washing with detergent will remove acid phosphatase, it has been shown that sufficient spermatozoa are retained in cotton fabric after washing such that a DNA profile can be obtained (Crowe et al., 2000). Consequently, it may be useful to sample areas of a possible seminal stain, such as the crotch of a pair of women's underpants, regardless of whether the item has been washed. It has also been shown that spermatozoa may be detected on items previously washed with a semen-stained garment (Kafarowski et al., 1996). Thus, the presence of spermatozoa on undergarments after laundering needs to be interpreted with care. When a DNA profile is obtained, one must establish all possible sources of the spermatozoa.

Table 7.2 shows an example of a laboratory form used in the examination of clothing from sexual offenses. Such forms serve as a useful guide and summary of the multiple tests used in the examination.

7.3 Saliva

Forensically significant amounts of saliva can be deposited on clothing or other items through biting, licking, kissing, or sucking the item. An adult produces between 750 and 1500 ml of saliva per day (Jones, 2005). Saliva contains high levels of amylase, buccal (mouth) cells, and skin cells. Over a decade ago it was found that DNA can be extracted from saliva deposited on postage stamps (Hopkins et al., 1994), and now obtaining DNA from saliva is a routine process. Saliva from the offender may support the victim's allegation

Table 7.2 Form for Examination of Clothing in Sex Offense Cases

Case number		Time since last intercourse (hrs)	
Police force		Condom worn	Yes/No/Not known
Case type		Victim injured	Yes/No/Not known
Medical form	Yes/No	Victim menstruating	Yes/No/Not known
Time between incident and medical		Victim washed/bathed	Yes/No/Not known
Examination (hrs)		Victim toilet	Yes/No/Not known

Swab type	Underpants	Bra	Underclothing	Top	Trousers	Bedding	Other
Blood							
AP +/−							
AP reaction time							
Microscopy score							
Tails present							
NECs[1] present							
Bacteria/yeast present							
Proteinase K							
Choline result							
PSA result							
Saliva							
Sample submitted to DNA?							
Other comments							

[1] nucleated epithelial cells

of kissing, licking, or sucking. Human bite marks are most often found in sexually motivated violent crimes, but the criteria for bite mark comparison are still being developed and are unevenly applied. The accompanying saliva, however, may be DNA profiled. The following case study (Schiro et al., 2007) describes crucial saliva evidence from clothing that helped overturn a conviction based on flawed bite mark evidence:

> A bar attendant was found murdered at her workplace in 1991 in Phoenix, Arizona. She was stabbed eleven times and bitten twice — once on her left breast through the tank top that she was wearing and once on her neck. Ray Krone was convicted of her murder and sentenced to death, due principally to bite mark evidence. It was alleged that the bite marks matched the dentition of Krone, and he was dubbed the "snaggletooth killer" due to his misshapen tooth. As a result of further investigation by his lawyer, it was discovered that saliva was present on the tank top in the area of the bite mark. The saliva was DNA profiled and excluded Krone; eventually, it matched on the CODIS database a convicted inmate of the Arizona prison system, Kenneth Phillips. Krone was released in 2002, and Phillips entered a guilty plea in 2006.

Saliva stains are difficult to locate because they are colorless, although sometimes they dry to a whitish stain. It is important to receive as much information as possible regarding the crime scenario. Any clothing that has allegedly been kissed or bitten or clothing that has covered areas of the victim's body that has had saliva deposited from the offender may be searched for saliva. Such clothing could include scarves, collars of sweaters or blouses, and the inside of brassiere cups or underpants.

Saliva may fluoresce under ultraviolet light (see Chapter 3 for assisted light screening techniques). The examiner must know the suspected details of the assault in order to consider the possible location of the saliva, so that screening tests can be employed efficiently.

7.3.1 Testing for Saliva

There is no test specific for saliva. Presumptive chemical tests rely primarily on the detection of alpha amylase. This enzyme is found in saliva up to 50 times higher than in other body fluids. The Phadebas® reagent is used in a press test and relies on a color reaction (James and Nordby, 2005; Myers and Adkins, 2008). The reagent is also used in a tube test. The Polilight® (alternate light source) is comparable in terms of sensitivity to the Phadebas® test, and DNA could be subsequently successfully recovered (Vandenberg and van Oorschot, 2006). Consequently, this light-assisted technique may be preferable to the Phadebas® test when there is minimal sample suspected.

Samples of the saliva stain may be analyzed for DNA. The examiner should also submit a control sample of an unstained area for DNA profiling at the same time. Because no confirmatory test exists for saliva, it is not possible to confirm a link between the presumptive, case-indicative presence of saliva and a DNA result. Other body fluids should be considered as a source of the DNA and case scenarios and hypothesis testing employed to determine whether the saliva–DNA link is the best explanation for the results.

7.4 Vaginal Secretions, Urine, Feces, and Vomit

7.4.1 Vaginal Secretions

Vaginal secretions are whitish or creamy stains, sometimes with a typical odor, and are generally found on the inner crotch of an adult female's underpants. Tests for vaginal secretions on clothing are not usually performed, because their presence does not necessarily indicate criminal activity. Generally, it is more important to sample the suspect stain for DNA analysis that could lead to a link with an individual. For example, the examiner might want to test for suspected vaginal drainage on bedclothes or other fabrics if a suspect claims that the victim had not been there. Vaginal secretions on objects that may have been inserted into the vagina, such as bottles, could also support a victim's story.

No specific confirmatory tests exist for vaginal secretions. Epithelial cells, which are glycogenated, will indicate they are from the vagina, but these cells are not unique to the vagina and may be found in smaller amounts in the mouth, for example. It is currently not possible to distinguish specifically between skin, buccal, vaginal, and penile epithelial cells using either cytological approaches or DNA profiling techniques (Paterson et al., 2006).

7.4.2 Urine

Urine is composed mainly of water and salts and has a characteristic odor and a light-yellow color. A urine stain on a garment cannot be located simply by odor because it permeates the whole garment and will not be localized to the urine stain. A urine stain may be located by its light yellowish color, but this is difficult if the background color of the garment is a similar color or is dark. Sometimes a slight change in luster on the garment is the only visual indication of a urine stain.

A finding of urine may be of interest in corroborating or casting doubt on a victim's or suspect's account of events. However, analysis of suspect urine stains are infrequently performed due to the insensitivity of the tests. The chemical tests for urine rely on the presence of urea and creatinine, but they are not specific because these substances are also found in other body fluids. Some cellular content from epithelial cells may be washed from the urinary tract, but urine stains are dilute and it is unlikely a DNA profile will be obtained from a typical urine stain on clothing. Nevertheless, there have been recent reports of success (Nakazono et al., 2005; Nakazono et al., 2008).

7.4.3 Feces

Human feces consist of undigested and digested food residue, mucosal cells, and bacteria. Fecal stains are detected on clothing by their characteristic appearance and odor. Their microscopic appearance will show undigested and partly digested remnants of plant material, such as cell walls, starch grains, and striated muscle from meat.

A chemical test for feces is the detection of urobilinoids, the final products of red blood cell breakdown, but this is not confirmatory. Although cells from the gastrointestinal tract will end up in fecal matter, DNA analysis has generally not been successful due to the large amounts of bacteria and digestive enzymes that degrade the DNA. Again, some researchers have reported success in obtaining DNA profiles (Vandenberg and van Oorschot, 2002).

7.4.4 Vomit

Deposits of vomit on clothing have the appearance of food stains: fibrous and particulate materials in a dried liquid matrix. There is no specific chemical test for the identification of vomit, although it has a characteristic appearance and odor. Macroscopic and microscopic identification of food is the most common technique.

7.5 Dandruff

Dandruff is essentially flakes of skin from the scalp and is a common material found in headwear. Dandruff can identify the usual wearer of a garment or a person who has recently worn it. The evidential value of dandruff has increased with the increased sensitivity of DNA analysis. DNA may be obtained from any nucleated epithelial cells found in dandruff, and it has been estimated that each dandruff particle may contain from 0.8 to 16.6 ng of DNA (Herber and Herold, 1998). Dandruff may also be found around the collars and shoulders of upper body garments. Medical conditions such as psoriasis and seborrheic dermatitis will also cause skin deposits in a manner similar to dandruff but on areas inside clothing associated with the problem. Dandruff is still a problem among the normal healthy population, with estimates that about 20% of the population suffer from the condition (Herber and Herold, 1998).

Most dandruff particles appear whitish and dry, although they may be yellowish and sticky in the presence of sebum (found in cases of seborrheic dermatitis, for example). The particles may be relatively large, 1 to 2 mm in diameter and easily visible by the naked eye, or about 0.5 mm in diameter and harder in consistency (Herber and Herold, 1998). A visual screen with the naked eye or with a stereomicroscope is usually sufficient to detect dandruff, especially if the clothing material is dark in color.

Blood, semen, or other biological material on a clothing item may mask the presence of dandruff. It is not uncommon to obtain mixtures from hats or balaclavas as a result of their being worn by more than one individual. Thus, we recommend collecting a number of dandruff particles from a garment and collecting single dandruff particles separately for potential DNA analysis to elucidate any potential mixtures.

7.6 DNA

Today most people understand that it is DNA that makes every individual unique. DNA resides in every cell of the body with the exception of red blood cells. The white cells in blood carry DNA in their nucleus; mature red blood cells have no nucleus and thus no DNA. The term "DNA" without a prefix generally implies nuclear (in the nucleus of a cell) DNA.

Prior to DNA typing, conventional genetic markers were used on physiological fluids such as blood and semen. These markers used the theories of population genetics that DNA analysis also uses, although the resulting frequencies of occurrence did not permit individualization. ABO grouping was one marker understood by the general public because most people know their ABO group. Also known as serology markers, they included protein and enzyme markers. Conventional serology tests, lacking the discriminating power of DNA typing, have now been superseded in most forensic laboratories, apart from initial exclusionary tests.

DNA profiling was first introduced into the forensic arena in the mid-1980s in England, with some noted journal publications (Jeffreys et al., 1985; Gill et al., 1985) and a high-profile crime investigation (Wambaugh, 1989). Since then, there has been an explosion in the use of DNA techniques in criminal investigation. Current methods of DNA typing use STRs, which are short unit repeat loci in the DNA molecule. A reasonable number of STR loci chosen (such as 9 or more) provide a high level of individualization in the population chosen for the sample. STR markers have become important tools for human identity testing and will continue to be used for many years because of their high degree of variability, ease of use in multiple amplifications, and implementation in national DNA databases; a core set of STR loci provides the capability of national and international sharing of criminal DNA profiles (Butler, 2006).

Once a biological sample has been located on a clothing item, the following steps are required for STR analysis:

1. Extraction of DNA from the biological sample
2. Quantification of the obtained DNA
3. Amplification of STR loci by polymerase chain reaction (PCR)
4. Separation and detection of amplification products by capillary electrophoresis

Accredited forensic laboratories will have their own validated DNA systems. The scientific working group SWGDAM has provided guidelines for the analysis and interpretation of DNA in forensic cases (SWGDAM, 2000; SWGDAM, 2001). There are numerous peer-reviewed articles, books, and reports on DNA analysis, suitable as references for the clothing examiner (e.g., Butler, 2005; National Research Council, 1996; Rudin and Inman, 2001).

Dirt, grease, and other substances present on or in clothing may either inhibit the PCR reaction or degrade any DNA present, thus limiting the amount of DNA obtained from the biological evidence. The examiner should carefully note the condition of the biological fluid on the garment and whether a stain is partly or completely covered in dirt or other substance. Sampling of the stain for DNA analysis should avoid any such contaminants if at all possible.

National and regional DNA databases are in use in many countries through statutory legislation, including England, Canada, Australia, and the United States. These databases have provided many cold "hits" for unknown samples left on clothing. Thus, in addition to comparison with provided reference samples, biological evidential samples from clothing may also be compared to DNA databases to find a potential "match" with a convicted offender.

Other DNA techniques in criminal cases that use different sequences are mitochondrial (mt) DNA and Y-STR profiling. Mitochondrial DNA is found in the mitochondria of cells and is maternally inherited, unlike nuclear DNA, which is biparentally inherited. It also has lower discrimination power compared to "regular" DNA typing. Mitochondrial DNA is useful for degraded samples, "ancient" samples (such as very old bones), and hair shafts. The technique is available only in very large forensic laboratories due to its expense. A large literature base exists for mtDNA (e.g., Holland and Parsons, 1999; Isenberg, 2005; SWGDAM, 2003; Tully et al., 2001).

Y-STR profiling analyzes the variation on the male (Y) chromosome in nuclear DNA. The Y chromosome is paternally inherited. This technique is useful when there are low levels of male DNA and high background levels of female DNA, in mixtures where the female portion is present in overwhelming quantities compared to the male portion, or where there are multiple male contributors. Examples of types of biological evidence on clothing for which this may be useful include admixed male and female bloodstains on the clothing of a deceased female who has bled copiously or analysis of a semen stain in a male gang rape case of a single female victim. The literature on this topic is growing (Daniels et al., 2004; Gill et al., 2006; Gusmao et al., 2006; Shewale et al., 2003).

Interpretation of nuclear DNA profiles and their accompanying statistics is also covered extensively in the literature (e.g., Balding, 2005; Buckleton et al., 2004; Clayton et al., 1998, Taroni et al., 2007; Triggs et al., 2000). If a DNA profile of a bloodstain on a T-shirt apparently matches a reference sample DNA profile, then what is the significance of that match? Two alternative population genetics models are used: the "product rule," which includes the assumption of Hardy-Weinberg and linkage equilibrium, and the "drift model," which incorporates a correction for population subdivision (Buckleton et al., 2001). The probability of a particular DNA profile match in the relevant population is calculated in England and Australia using Bayes' theorem and likelihood ratios (Foreman and Evett, 2001). Forensic laboratories in the United States predominantly use random match probability and the product rule (National Research Council, 1996; Budowle et al., 2000). There has been debate in the literature as to which of these statistical methods are appropriate (Buckleton et al., 2001; Weir, 2001). Clothing examiners who report DNA results should be aware of these major issues.

DNA statistics has become such a popular topic that it now occupies chapters in general forensic statistics texts (Aitken and Taroni, 2004). Furthermore, the acceptance of DNA and its accompanying statistics in criminal trials has changed the way the legal system views forensic evidence in general. The legal system now appears to expect statistical values placed on nonbiological forensic evidence, whether applicable or not (Taupin, 2004). (Further discussion of statistics for the clothing examiner is provided in Chapter 9.)

7.7 Wearer DNA

We use the term "wearer DNA" for DNA that has been deposited on clothing by the wearer of that clothing. This DNA has been deposited through contact with the skin during the wearing of the clothing and consists of nucleated epithelial cells. The single "habitual" wearer of the garment should be detected as the major source of wearer DNA on that garment. Minor DNA profiles may also be detected from individuals with whom the habitual wearer has had close contact or from someone who has borrowed the garment.

Nucleated cells from other body areas, such as the eyes, nose, or mouth, also yield successful DNA profiles. The hands may act as vectors of transmission of these cells to different parts of the clothing. Clothing that may yield successful DNA profiles due to regular contact with the skin or with mouth and nose secretions include socks, gloves, baseball caps (sweat band), balaclavas, shirt underarms, shoelaces, and the inside edge of the fly of underpants (Wickenheiser, 2002). Even lipstick has been found to transfer sufficient DNA from the "wearer" of the lipstick (Webb et al., 2001), so lipstick found on clothing is a further avenue of DNA analysis.

The DNA profile of the wearer can also be obtained from shoes, most notably from the insoles (Bright and Petricevic, 2004). The area in contact with the top of the foot may also prove feasible for sampling in suitable shoes. Bedding such as sheets may yield sufficient DNA from the habitual sleeper, but it has also been noted that a DNA profile of an individual can be obtained from bedding after just one night of sleep (Petricevic et al., 2006). These results have implications for clothing other than shoes because they indicate the more an item is worn without cleaning, the greater the amount of DNA that can be sampled. Shoe insoles may be worn for months or more without cleaning and may contain a reservoir of DNA. Bedding has close contact with the skin overnight and if sheets are not changed regularly, they may also contain a reservoir of DNA. If an item of clothing appears grubby and well worn, then the potential for DNA of the habitual wearer of the clothing is high.

7.8 Trace DNA

We define trace DNA found on a garment as low levels of DNA whose cell type cannot be identified. It also cannot be stated when these cells were deposited. Blood and semen can be more readily linked with a crime due to their association with violent or intimate contact. However, trace DNA from an unspecified cellular source reduces this relevance of biological evidence to a crime. There is an increase in the degree of uncertainty as to how the DNA may have been transferred to the garment, and the relevance of findings may be difficult to assess. In fact, the emphasis in the ultimate court case may shift from "Whose DNA is this?" to "How did this person's DNA get here?" (Evett et al., 2002). The standard considerations of any type of trace evidence, such as transfer, persistence, and recovery, will also be important.

It has been shown that simply touching an object will leave sufficient amounts of DNA for a profile (van Oorschot and Jones, 1997). It has also been recognized that some individuals may have an increased propensity to shed DNA-containing cells compared to the rest of the population; these people are called "shedders" (Lowe et al., 2002). Knowledge of an individual's shedding characteristics may be useful for general background data in the interpretation of DNA trace evidence (Phipps and Petricevic, 2007).

Care should be taken in speculating the somatic origin of DNA-bearing cells found on clothing where trace DNA profiles have been obtained but the originating source has not been determined. Minute undetected quantities of blood, saliva, semen, and even mixtures can be the origin of the DNA profiles, as well as "wearer" DNA and skin cells. Replicate samples from different areas of the garment should be considered.

7.9 Multiple Body Sources

The amount of DNA per volume of sample material or exhibit varies depending on the source. "Solid" samples of DNA such as human tissue bear very large amounts of DNA per unit volume, as do sperm. These samples should be considered as having the highest DNA potential when there are two or more types of DNA-bearing cells on a garment (Wickenheiser, 2002). Blood has the next-highest DNA potential because, although blood is prevalent in violent crimes, the DNA-bearing white blood cells are outnumbered 400 to 1 by the red blood cells. Saliva and nose and mouth secretions contain the third-highest DNA potential because of the small volume of body fluid conveying the DNA-bearing cells and the small contact area. Given the transient nature, trace or wearer DNA has the lowest DNA potential (Wickenheiser, 2002).

Spermatozoa are often commingled with cells (such as epithelial cells) from the alleged victim, especially on garments that were worn at the time of the alleged offense. Complete separation using differential extraction is the aim, but it is not always successful. A mixed DNA profile with a female contribution indicates the presence of female cells in such a situation.

An estimate of the quantity of DNA present on the garment is useful to assist in the interpretation of the relevance of a DNA profile. If a visible fresh bloodstain yields several micrograms of DNA, then it is reasonable to associate the obtained DNA profile with the bloodstain. However, the association would be uncertain if the bloodstain was minute, old, and yielded just a few picograms of DNA.

The association of body fluids and DNA profiles is thus not implicit. A small, degraded blood spot that has given a positive presumptive test for blood might be masked by a fresh saliva stain that contributes to the observed result. Thus the examiner cannot infer either the type of cell donating the DNA or the time when the cells were deposited from only a simple examination.

Evidence types on clothing that are not discrete entities (for example, bloodstains as opposed to a single uncoated hair) need to be considered especially. Chances are increased that a DNA profile may not be directly related to the evidential body fluid that is supposedly analyzed.

Table 7.3 presents a suggested order of removal of biological matter when more than one type is present, according to body origin (Wickenheiser, 2002).

The examiner must consider relative levels of DNA potential in order to minimize the potential masking of DNA profiles. Light smudges of suspect blood on the outside of a windjacket may be masked by large amounts of DNA from the wearer of the clothing if a sample of the garment is taken and analyzed without regard for the body origin of the DNA. The blood smear should be examined under the stereomicroscope to determine the extent of the blood and the stained yarns/material carefully sampled by shaving or cutting. If wearer DNA is the target material, such as from a bloodied balaclava dropped at a

Table 7.3 Suggested Order of Removal

DNA potential	1	2	3	4
Body origin	Tissue, semen	Blood	Saliva, nose, mouth	TraceDNA
Order of removal	Last	Third	Second	First

Reprinted, with permission, and adapted from Wickenheiser, R.A., Journal of Forensic Sciences, Volume 47(3), 2002, copyright ASTM International, 100 Barr Harbor Drive, West Conshohocken, PA 19428.

murder scene, the inside surfaces of the balaclava should be examined under the stereomicroscope to determine unstained areas that would be in contact with the skin, nose, or mouth. These unstained surfaces provide the potential for wearer DNA. The blood on the outside surface presumably belongs to the victim — although not always; the examiner should formulate meaningful hypotheses for the crime scenario.

The examiner should also consider the order in which samples are taken from the garment. High-yielding DNA sources should be analyzed last versus trace DNA sources, although this can pose logistical problems and may be unworkable. Any collection method that dislodges or distributes the high-yielding DNA source on a garment may mask the minor DNA source (such as wearer DNA) in a mixture, so a sampling rationale is paramount. The item of clothing should be handled so that higher yielding and different sources of DNA are not touched and then transferred to the lower yielding sources of DNA during examination.

It may be justified to thus sample multiple areas of the same stain, or sample multiple stains, on the garment in order to elucidate more than one contributor or more than one body source. As always, sampling needs to be considered in terms of the case as a whole, together with hypotheses formulated.

7.10 Mixtures

Mixtures from different body fluid sources and/or different individuals may occur on one garment and in one location or stain. It may be possible to localize a particular stain to a biological fluid; for example, a bloodstain may have only partly encroached upon a semen stain. The two stains may then be analyzed separately, although until a DNA result is obtained it will not be known whether there was successful separation of the material from different individuals. A DNA profile will not inform the examiner as to body fluid source if it is from one individual. If the stains are reasonably large or sufficient in number, then the examiner may sample a variety of stains without consuming limited valuable evidence.

The interpretation of DNA profiles from more than one contributor is one of the most complex activities facing a clothing examiner who reports DNA profiles. Laboratory guidelines and the literature have attempted to provide a body of theory for a coherent treatment of mixed stains (Weir et al., 1997; Clayton et al., 1998; Curran et al., 1999; Gill et al., 2006). Mixtures with one major contributor and one minor contributor are the simplest to analyze, especially when there is a known contributor (victim). More problematic is when a major contributor cannot be determined or there are more than two contributors.

A mixed DNA profile from two individuals and two possible body sources is also problematic. As an example, the analysis of semen staining on a complainant's bed sheet may yield a mixed DNA profile. The major DNA component (or sperm fraction from a differential lysis) of a stain, likely to be from semen, corresponds to a suspect and the minor component, likely to be from epithelial cells, corresponds to the complainant. It has been considered (Petricevic et al., 2006) that it is not possible in this situation to determine whether the mixed stain occurred from a mixing of biological fluids during sexual intercourse. Nor is it possible to determine whether the DNA was deposited at the same time. In the case of sequential deposits, sampling the margins of a stain may elucidate the components, especially if the stain is asymmetrical or gives other indications of possible separate deposits (Cwiklik and Gleim, 2009).

7.11 Nonhuman Biological Evidence

A discussion of animal and plant biological evidence is beyond the scope of this chapter and book, but the forensic clothing examiner should be aware of this type of material. Blood and other body fluids from animals are sometimes found on clothing and may be important to a case. There is a body of literature on this topic and a growing forensic veterinary literature. DNA testing of animal tissue, although not as specific as testing of human tissue, has been reported in the literature and used in casework. Although most plant material encountered on clothing can be tested by microscopy, if it is important to establish a particular strain of plant material, DNA testing may be of value.

7.12 Conclusion

Examining and reporting human biological evidence on clothing is one of the more challenging tasks facing the forensic examiner, especially with the advent of DNA profiling. Methods for locating and sampling biological stains are essential to the successful interpretation of DNA analysis and identification of body fluids, especially when mixtures are involved. A choice of samples based on potential significance and case hypotheses is necessary.

If the examiner also reports the DNA result, then that examiner is required and often tested by the court system to be up to date with the DNA literature, a continual growth area. With the support of their laboratory, examiners should be confident in their examination and their interpretation of this valuable evidence.

New methods for detecting, preserving, and analyzing DNA are continually being developed. These methods may also aid in recovery of DNA from biological evidence. Different nuclear markers may also be discovered. These new techniques may help the examiner to detect even lower levels of DNA on clothing or simplify their current examination. Furthermore, genetic techniques for the definitive identification of body fluids are being developed. A messenger RNA method for the definitive identification of blood, saliva, semen, and vaginal secretions (even in mixed stains) using selected genes expressed in a tissue-specific manner has been described (Juusola and Ballantyne, 2005). These new techniques have the potential to replace the labor-intensive battery of physical and chemical tests described in this chapter.

Nevertheless, with these newer techniques it will remain important for the examiner to preserve the integrity of the evidence once received, especially to prevent contamination. Strict protocols should be followed regarding health and safety, prevention of contamination, and preservation of the evidence for further or possible future examination.

7.13 References

Adair, T.W. and Shaw, R.L., Enhancement of bloodstains on washed clothing using luminol and LCV reagents, *IABPA News*, December 2005.

Aitken, C.G.G. and Taroni, F., *Statistics and the Evaluation of Evidence for Forensic Scientists*, 2nd ed., John Wiley & Sons, Chichester, England, 2004.

Balding, D., *Weight-of-Evidence for Forensic DNA Profiles*, John Wiley & Sons, Chichester, England, 2005.

Bright, J.A. and Petricevic, S.F., Recovery of trace DNA and its application to DNA profiling of shoe insoles, *Forensic Sci. Int.*, 145, 7–12, 2004.

Buckleton, J.S., Walsh, S., and Harbison, S.A., The fallacy of independence testing and the use of the product rule, *Sci. Justice*, 41, 81–84, 2001.

Buckleton, J.S., Triggs, C.M., and Walsh, S.J., Eds., *Forensic DNA Evidence Interpretation*, CRC Press, Boca Raton, FL, 2004.

Budowle, B., Chakraborty, R., Carmody G., et al., Source allocation of a forensic DNA profile, *Forensic Sci. Commun.*, 2(3), July 2000.

Butler, J.M., *Forensic DNA Typing: Biology, Technology and Genetics of STR Markers*, 2nd ed., Elsevier Academic Press, Burlington, MA, 2005.

Butler, J.M., Genetics and genomics of core short tandem repeat loci used in human identity testing, *J. Forensic Sci.*, 51(2), 253–265, 2006.

Clayton, T., Whitaker, J.P., and Sparkes, R., Analysis and interpretation of mixed forensic stains using DNA STR profiling, *Forensic Sci. Int.*, 91(1), 55–70, 1998.

Crowe, G., Moss D., and Elliot, D., The effect of laundering on the detection of acid phosphatase and spermatozoa on cotton T-shirts, *Can. Soc. Forensic Sci. J.*, 33, 1–5, 2000.

Curran, J.M., Triggs, C.M., Buckleton, J.S., et al., Interpreting DNA mixtures in structured populations, *J. Forensic Sci.*, 44(5), 987–995, 1999.

Cwiklik, C.L. and Gleim, K.G., Forensic casework from start to finish, in *Forensic Science Handbook*, Vol. III, pp. 1–30, Saferstein, R., Ed., Prentice-Hall, Englewood Cliffs, NJ, 2009.

Daniels, D.L., Hall, A.M., and Ballantyne, J., SWGDAM developmental validation of a 19-locus Y-STR system for forensic casework, *J. Forensic Sci.*, 49(4), 668–683, 2004.

Evett, I.W., Gill, P., Jackson, G.M., et al., Interpreting small quantities of DNA: The hierarchy of propositions and the use of Bayesian networks, *J. Forensic Sci.*, 47(3), 520–530, 2002.

Foreman, L.A. and Evett, I.W., Statistical analyses to support forensic interpretation for a new ten-locus STR profiling system, *Int. J. Legal Med.*, 114, 147–155, 2001.

Gaensslen, R.E., *Sourcebook in Forensic Serology, Immunology and Biochemistry*, U.S. Government Printing Office, Washington, DC, 1983.

Gill, P., Jeffreys, A.J., and Werrett, D.J., Forensic application of DNA "fingerprints," *Nature*, 318, 577–579, 1985.

Gill, P., Brenner, C.H., Buckleton, J.S., et al., DNA Commission of the International Society of Forensic Genetics: Recommendations on the interpretation of mixtures, *Forensic Sci. Int.*, 160, 90–101, 2006.

Gusmao, L., Butler, J.M., Carracedo, A., et al., DNA Commission of the International Society of Forensic Genetics (ISFG): An update of the recommendation on the use of Y-STRs in forensic analysis, *Forensic Sci. Int.*, 157, 187–197, 2006.

Herber, B. and Herold, K., DNA typing of human dandruff, *J. Forensic Sci.*, 43(3), 648–656, 1998.

Hochmeister, M.N., Borer, U., Budowle, B., et al., Evaluation of prostate-specific antigen (PSA) membrane test assays for the forensic identification of seminal fluid, *J. Forensic Sci.*, 44, 1057–1060, 1999.

Holland, M.M. and Parsons, T.J., Mitochondrial DNA sequence analysis: Validation and use for forensic casework, *Forensic Sci. Rev.*, 11, 22–50, 1999.

Hopkins, B., Williams, N.J., Webb, M.B.T., Debenham, P.G., and Jeffreys, A.J., The use of minisatellite variant repeat polymerase chain reaction (MV-PCR) to determine the source of saliva on a used postage stamp, *J. Forensic Sci.*, 39(2), 526–531, 1994.

Isenberg, A.R., Forensic mitochondrial DNA analysis, in *Forensic Science Handbook*, Saferstein R., Ed., Vol. II, 2nd ed., Pearson Prentice-Hall, Upper Saddle River, NJ, 2005.
James, S.H. and Nordby, J.J., Eds., *Forensic Science: An Introduction to Scientific and Investigative Techniques*, 2nd ed., Taylor & Francis, Boca Raton, FL, 2005.
Jeffreys, A., Wilson, V., and Thein, S.L., Individual specific "fingerprints" of human DNA, *Nature*, 316, 76–79, 1985.
Jones, E.L., Jr., The identification of semen and other body fluids, in *Forensic Science Handbook*, Saferstein R., Ed., Vol. II, 2nd ed., Pearson Prentice-Hall, Upper Saddle River, NJ, 2005.
Juusola, J. and Ballantyne, J., Multiplex mRNA profiling for the identification of body fluids, *Forensic Sci. Int.*, 152, 1–12, 2005.
Kafarowski, E., Lyon, A.M., and Sloan, M.M., The retention and transfer of spermatozoa in clothing by machine washing, *Can. Soc. Forensic Sci. J.*, 29(1), 7–11, 1996.
Khaldi, N., Miras, A., and Gromb, S., Evaluation of three rapid detection methods for the forensic identification of seminal fluid in rape cases, *J. Forensic Sci.*, 49(4), 1–5, 2004.
Kobus, H.J., Silenieks, E., and Scharnberg, J., Improving the effectiveness of fluorescence for the detection of semen stains on fabrics, *J. Forensic Sci.*, 47(4), 819–823, 2002.
Lowe, A., Murray, C., Whitaker, J., et al., The propensity of individuals to deposit DNA and secondary transfer of low level DNA from individuals to inert surfaces, *Forensic Sci. Int.*, 129, 25–34, 2002.
Manna, A.D. and Montpetit, S., A novel approach to obtaining reliable PCR results from luminol treated bloodstains, *J. Forensic Sci.*, 45(4), 886–890, 2000.
Myers, J.R. and Adkins, W.K., Comparison of modern techniques for saliva screening, *J. Forensic Sci.*, 53(4), 862–867, 2008.
Nakazono, T., Kashimura, S., Hayashiba, Y., et al., Successful DNA typing of urine stains using a DNA purification kit following diafiltration, *J. Forensic Sci.*, 50(4), 860–864, 2005.
Nakazono, T., Kashimura, S., Hayashibia, Y., et al., Dual examinations for identification of urine as being of human origin and for DNA-typing from small stains of human urine, *J. Forensic Sci.*, 53(2), 359–363, 2008.
National Research Council, The evaluation of forensic DNA evidence, National Academies Press, Washington, D.C., 1996.
Paterson, S.K., Jensen, C.G., Vintiner, S.K, et al., Immunohistochemical staining as a potential method for the identification of vaginal epithelial cells in forensic casework, *J. Forensic Sci.*, 51(5), 1138–1143, 2006.
Petricevic, S.F., Bright, J.A., and Cockerton, S.L., DNA profiling of trace DNA recovered from bedding, *Forensic Sci. Int.*, 159, 21–26, 2006.
Pex, J.O., The use and limitations of luminol in bloodstain pattern analysis, *IABPA News*, December 2005.
Phipps, M. and Petricevic, S.F., The tendency of individuals to transfer DNA to handled items, *Forensic Sci. Int.*, 168(2–3), 162–168, 2007.
Rawley, A. and Caddy, B., *Damilola Taylor: An Independent Review of Forensic Examination of Evidence by the Forensic Science Service*, Home Office, United Kingdom, April 2007. Available at http://homeoffice.gov.uk; accessed April 2008.
Reynolds, M., The ABAcard® Hematrace® — A confirmatory identification of human blood located at crime scenes, *IABPA News*, June 2004.
Rudin, N. and Inman, K., *An Introduction to Forensic DNA Analysis*, 2nd ed., CRC Press, Boca Raton, FL, 2001.
Sanders, C.T., Sanchez, N., and Ballantyne, J., et al., Laser microdissection separation of pure spermatozoa from epithelial cells for short tandem repeat analysis, *J. Forensic Sci.*, 51(4), 748–757, 2006.
Schiro, G.J., Streed, T.B., Barham, E.T., et al., Anatomy of a wrongful conviction: A multidisciplinary examination of the Ray Krone case, *Proceedings of the American Academy of Forensic Sciences*, Vol. XIII, February 2007. Available at http://aafs.org/pdf/07Proceedings.Complete.pdf/; accessed April 2008.

Scientific Working Group on DNA Analysis Methods (SWGDAM), Short tandem repeat (STR) interpretation guidelines, *Forensic Sci. Commun.*, 2(3), July 2000.
Scientific Working Group on DNA Analysis Methods (SWGDAM), Training guidelines, *Forensic Sci. Commun.*, 3(4), October 2001.
Scientific Working Group on DNA Analysis Methods (SWGDAM), Guidelines for mitochondrial DNA (mt DNA) nucleotide sequence interpretation, *Forensic Sci. Commun.*, 5(2), April 2003.
Shaler, R.C., Modern forensic biology, in *Forensic Science Handbook*, Saferstein R., Ed., Vol. II 2nd ed., Prentice-Hall, Upper Saddle River, NJ, 2002.
Shewale, J.G., Sikka, S.C., Schneida, E., and Sinha, S.K., DNA profiling of azoospermic semen samples from vasectomised males by using Y-PLEX™6 amplification kit, *J. Forensic Sci.*, 48(1), 127–129, 2003.
Soares-Vieira, J.A., Billerbeck, A.C.E., and Iwamura, E.S.M., Y-STR's in forensic medicine: DNA analysis on semen samples of azoospermic individuals, *J. Forensic Sci.*, 52(3), 664–670, 2007.
Taroni, F., Bozza, S., Bernard, M., et al., Value of DNA tests: A decision perspective, *J. Forensic Sci.*, 52(1), 31–39, 2007.
Taupin, J.M., Forensic hair morphology comparison — A dying art or junk science? *Sci. Justice*, 44(2), 95–100, 2004.
Tobe, S.S., Watson, N., and Daeid, N.N., Evaluation of six presumptive tests for blood, their specificity, sensitivity, and effect on high molecular-weight DNA, *J. Forensic Sci.*, 52(1), 102–109, 2007.
Triggs, C., Harbison, S.A., and Buckleton, J., The calculation of DNA match probabilities in mixed race populations, *Sci. Justice*, 40, 33–38, 2000.
Tully, G., Bar, W., Brinkmann, B., et al., Considerations by the European DNA profiling (EDNAP) group on the working practices, nomenclature, and interpretation of mt DNA profiles, *Forensic Sci. Int.*, 124, 83–91, 2001.
van Oorschot, R.A.H. and Jones, M.K., DNA fingerprints from fingerprints, *Nature*, 387, 767, 1997.
Vandenberg, N. and van Oorschot, R.A.H., Extraction of human nuclear DNA from faeces samples using the QIAamp DNA Stool Mini Kit, *J. Forensic Sci.*, 47, 993–995, 2002.
Vandenberg, N. and van Oorschot, R.A.H., The use of Polilight® in the detection of seminal fluid, saliva, and bloodstains and comparison with conventional chemical-based screening tests, *J. Forensic Sci.*, 51(2), 361–370, 2006.
Wambaugh, J., *The Blooding*, William Morrow, New York, 1989.
Webb, L.G., Egan, S.E., and Turbett, G.R., Recovery of DNA for forensic analysis from lip cosmetics, *J. Forensic Sci.*, 46(6), 1474–1479, 2001.
Weir, B.S., DNA match and profile probabilities: Comment on Budowle et al. (2000) and Fung and Hu (2000), *Forensic Sci. Commun.*, January 2001.
Weir, B.S., Triggs, C.M., and Starling, L., Interpreting DNA mixtures, *J. Forensic Sci.*, 42(2), 213–222, 1997.
Wickenheiser, R.A., Trace DNA: A review, discussion of theory, and application of the transfer of trace quantities of DNA through skin contact, *J. Forensic Sci.*, 47(3), 442–450, 2002.

Traces and Debris

8

When two objects or beings come into contact, a transfer of material occurs. This is commonly known as Locard's theorem (Thorwald, 1967) or Locard's exchange principle. Trace evidence consists of the material left behind from such transfers and may include stains, deposits, and debris. Stains and deposits — absorbed, adsorbed, firmly or loosely bonded to the substrate — are discussed in Chapter 4. Trace DNA, sometimes called touch DNA, is reviewed in Chapter 7. In this chapter we wil cover debris (accumulated fibrous and particulate deposits that adhere to a substrate via static electricity or mechanical entanglement) and traces (loosely adhering residues).

Debris is the accumulation of particles and fibrous material (including hairs) that results from the processes occurring in an environment or around a person. Examples of debris on clothing include dust and lint. A trace is usually a small, even microscopic, transfer of a substance left on an object either by contact with another object, person, or animal or by projection from one object or being onto another. For example, if someone leans against a tree, traces of bark and the things that live on it may transfer to the person's clothing. Projected material is deposited from fragmentation, such as glass shattering or seed pods exploding. Traces and debris found on clothing form a continuum rather than two discrete categories and are thus combined in this chapter.

Traces and debris may include materials of many types and sources. The debris characteristic of an item of clothing includes microscopic flakes and crusts of dried body fluids that are the basis for "trace" DNA, traces of skin, and traces of resident DNA from the wearers of a garment. A subset of debris from firearms-related activities may be found on clothing, especially if the wearer of the clothing owns and uses firearms or associates with people who do. Gunpowder particles, primer residue, detritus from fired bullets such as microscopic fragments of lead and copper, microscopic beads formed by melted bullet lead, hollow black or brown spheres from black powder weapons, etc., may thus be observed.

The first person who examines the clothing, regardless of specialty, should be aware of evidence types other than that of his or her own expertise. Every time a garment is handled, some evidence is inevitably altered or dislodged. Thus, it is incumbent on the first examiner to document and safeguard the stains, deposits, debris, DNA-bearing tissue, and firearms residue for further examination. Every examiner should be alert to the significance of touch DNA and firearms residues as traces that may have transferred via indirect contact, and to traces of controlled substances, explosives residues, or a population of mold or insects on an item of clothing. Evaluation of traces and debris may require a study of the context in which they are deposited and include examinations of companion deposits.

Traces and debris also include:

- Fibers
- Human and animal hairs
- Flaked, chipped, and sprayed paint

- Glass
- Soil
- Paper and plastics
- Wood, pollen, and other botanicals such as seeds, spores, and husks

Detritus from all the materials of daily life eventually appear as debris: glitter; foam rubber; insect parts and sometimes the insects themselves; animal detritus, including scat and feathers; industrial and agricultural dusts; particles from woodwork, machining, sewing, automotive work, local cottage industries, and so on. Debris is produced by welding, soldering, heating, burning and incinerating, and other processes that alter the parent material and may be associated with damage to the items of clothing on which they are found.

The focus of this book is clothing examination. The focus of this chapter is the information one can obtain from traces and debris using the naked eye, a stereomicroscope, and a few preliminary microscopic and chemical tests. This provides the basis for selecting samples for further testing to identify materials and to find, collect, and interpret deposits of traces and debris.

8.1 The Nature of Debris

Every particle of debris is produced by a process: shedding, scraping, grinding, tearing, abrasion and rubbing, fracture and shattering, spraying, and so on. (See Section 8.14 for a detailed list.) Processes that occur repeatedly in a given environment result in accumulations of particles that are characteristic of that environment, whether they be fine (usually particulate debris such as dust), mostly fibrous (debris such as lint), or some combination. Debris can accumulate in place or can be transported and transferred from other debris-bearing objects or beings.

Debris that is generated in a particular place or is repeatedly deposited there is characteristic of that environment. Debris that is generated by a person (such as hair and skin flakes) and that reflects a person's habits and environments is characteristic of that individual. A person's characteristic debris is usually found on his or her clothing. "For the microscopically finest particles that cover our clothing and our bodies are the mute witnesses to each of our movements and encounters" (Locard, 1930). A one-time event can also contribute to debris (for example, a flour spill or glass from a broken window), and the debris may serve as a marker of that event and provide a reference point for a time line or sequence of events.

A residue or trace of an object or creature can supply information about the person, object, or event. Debris comprises an accumulation of traces. Debris is a record of processes; thus it is historical. Collections of broken pottery and the remains of ancient cooking fires are used by archaeologists to reconstruct aspects of ancient civilizations. So debris on clothing can assist with reconstruction of more contemporary incidents and address specific questions about those incidents, such as whether a particular person broke a particular window (Crutcher, 1978; Palenik, 1982a, 1982b, 1982c, 1983; Petraco, 1986; Chisum and Turvey, 2006).

8.2 Sorting Tools for Evaluating Traces and Debris

Sorting tools for classifying stains and deposits are discussed in Chapter 4. The questions that can be addressed via examination with a stereomicroscope will be repeated here as applicable to traces and debris: What is it? Where did it come from (material or origin)?

How was it deposited? How did it transfer to the evidence item? What happened to it afterwards? When was it deposited? Is it from this event or from something else?

In addition, the following questions are useful for evaluating traces and debris: Is it from the substrate material or from debris adhering to it? Is it normal or foreign debris? How did it get there and how did it stick? Was it deposited via direct or intermediary transfer? These topics are addressed in the sections that follow.

Sorting tools are further described in Appendix 4. Suggested process-based terminology for traces and debris is in Section 8.14. The sorting tools and terminology are not a rigid classification system. Each examiner should use those that are most useful.

8.3 Composition of Debris

8.3.1 Normal Debris vs. Foreign Debris

Normal debris is the set of particles characteristic of or "native" to a person, object, or place. These are the collection of particles usually found in a particular room or in a particular pocket or on a specific person's clothing or hair. The debris of a woodworking shop consists mostly of wood shavings and sawdust. The debris in a kitchen includes food particles. The debris in an auto body shop includes paint chips, paint spray, metal shavings, and grease. The debris in a house includes fibers and paint chips from objects in the house and hair from the inhabitants and frequent visitors. A smoker's pockets contain tobacco particles. A nibbler's pockets contain food particles. An auto mechanic's fingernails, until scrubbed, have grease beneath them. All of us, to a greater or lesser degree, have fibers on our clothing from other clothing articles, blankets, and other objects in our households, as well as hair from family members and any domestic animals. Finally, normal debris on clothing items includes fibers and fiber pills from the item itself that have detached from the parent fabric and may include atypical fibers that have been discolored, melted, or distorted.

Foreign debris comprises the particles on a person, object, or place that were produced by something other than the habitual activities of a person or the processes that normally take place in an environment. It is only by evaluating normal debris that the foreign debris — the usual subject of comparisons — can be given significance. It is foreign debris that is usually discussed when a transfer of trace evidence is at issue; normal debris then forms the context for the evaluation. Only clearly foreign material can provide evidence of contact with another object or person. Consider a particular type of sawdust found on the floor of a workshop thought to be a crime scene. Its presence on the clothing of a rape victim would have little significance if that type of sawdust is also found in the victim's usual environment. It might, however, be significant if absent in the victim's everyday environment.

It is important for the scientist to know whether the individuals represented by the clothing and reference samples submitted to the laboratory are related to each other and to the sites involved in the incidents under investigation. If a member of a victim's household is suspected of a crime, it would be of little value to search for fibers of that person's clothing on the victim and vice versa, unless either person's clothing had just been purchased. The examination instead would focus on whatever was inherent in the incident itself; for example, bloodshed, overturned dusting powder, splashed soup, or a ligature (Cwiklik and Gleim, 2009). The ligature may hold traces of debris and sweat that would provide a link with the assailant. The debris on the victim's clothing should not be ignored — someone other than the suspect may be the true assailant. However, debris on the suspect's clothing would be of no particular interest unless something inherent to the crime was found there. The significance of an absence of corresponding debris would depend on the case circumstances and the history of the garment.

Of course, sometimes material on an article of clothing is neither characteristic of the wearer nor related to the event under investigation. Such materials may be from previous or subsequent events or may substantiate an alternative account of what the individual was doing. It is important for the person examining the clothing to report apparently unrelated foreign materials to alert police, prosecution, and defense attorneys, who may be able to respond to the information, especially if they have information not known to the scientist.

CASE EXAMPLE 8.1

Normal debris and foreign debris; relevance to incident

A young woman was killed in her bedroom by strangulation, and her upstairs neighbor was suspected of the crime. There was little bloodshed. Fibers consistent with the victim's living room carpet and furniture at the scene were found on the neighbor's shoes, as were several hairs consistent with the victim's reference hair samples from autopsy. These types of fibers and hairs were also found in the normal debris of the victim's living room. Because the suspect had helped her move into her apartment several weeks earlier and had visited on at least two occasions, the fibers on his shoes were not evidence of his involvement or even of his presence on the day of the murder. However, fibers consistent with the sock used as a ligature — apparently from a laundry basket near her bed — and leather fragments like those of the victim's leather watchband (not a part of the living room debris) were significant as links with the murder. In addition, the population of lint, animal hairs, and foreign hairs on the suspect's sweatshirt were reflected in the foreign debris on the victim's nightgown, and very few particles and fibers on the sweatshirt were absent from the nightgown. The examiner reached a strong opinion that the sweatshirt and the nightgown were in contact. A fiber pill found at autopsy on the victim's buttocks reflected the fibers and cat hairs in fiber pills in the crotch of the suspect's pants. A foreign pubic hair on the victim's buttocks was correlated with the suspect's pubic hair. The evidence of contact between the garments, together with the fiber pill and the pubic hair, established that not only the garments but also the victim and suspect were in contact, and that contact involved an attempt at sexual activity (Cwiklik, 1999, Case Example 5)

Among the hairs found on the victim's nightgown were hairs attributable to the neighbor's spouse. Was the spouse implicated in the murder? Deposits of hairs attributable to the spouse were evaluated with respect to other foreign debris on the nightgown already found to correspond with debris on the sweatshirt. The number and somatic (body area) origin of the hairs attributable to the spouse reflected the proportion of similar hairs on the neighbor's sweatshirt. This indicated secondary transfer from the neighbor's clothing rather than direct transfer from the spouse.

CASE EXAMPLE 8.2

Foreign debris attributed to other activities; reference samples examined; experiments to simulate alternative hypothesis

A man was seen loading tires stolen from a construction firm into his pickup truck. Police arrested the man and a friend who was with him. Both the man with the pickup and the friend said the friend was uninvolved. However, the responding officer observed black dust on the friend's clothing that the officer suspected was from handling the tires. According to the friend, the dust was from burning a coil of copper wiring, which he intended to melt down and sell to a recycler. However, he too was arrested and his clothing sent to the laboratory. Black dust was observed on the front of the pants and jacket, some of it in multiple nonparallel linear deposits more suggestive of contact with burnt wire than of contact with tires. When the black dust deposits were examined under a stereomicroscope, a few tiny copper-colored particles were observed. Preliminary samples collected on clear sticky-tape were examined at higher magnification. The numerous black particles observed did not have the typical rolled appearance of rubber fragments, but because they were somewhat curled and elongated, additional samples were examined. Test rubbings were obtained of new tires and tires currently in use on a variety of vehicles, including large trucks. Investigators also obtained rubbings from tires stored at the construction company. The particles in the rubbings from different sources exhibited a range of variation, but none were like those on the clothing.

The friend said that in melting down the copper wiring, he stacked some scrap wood and then placed the coil of copper wiring on top, sandwiching it between newspapers that he used as kindling. The laboratory obtained a length of copper wiring of that type and of other types sold in the local building supply store. Several sections of each type were wrapped in newspaper, placed on pieces of wood, and burnt. The plastic coating of several types of wiring produced curled, elongated sooty particles like those observed on the clothing. The scientist planned to further analyze the particles using FTIR for organics and SEM-EDX for the presence of copper. Before these tests could be conducted, police obtained a report from a forest ranger who had issued a warning to the friend for burning copper wiring in a national forest. Did the friend burn the wiring and also help steal tires? The black dust included only a very few particles not eliminated as tire rubber by microscopic examination. If the friend had been handling tires, more than a few particles of rubber dust would be expected. The charge of stealing tires was dropped.

8.3.2 Individual Types of Material vs. Sets of Debris

When a burglar breaks a window and reaches in to open the latch, the shattered glass is of interest, as are any deposits of fibers, blood, or hair that may transfer to the broken edges of the window and any shoe prints or knee prints found in any adjacent soil. When examining the suspect's clothing, the scientist should pay particular attention to shards of glass, to glass cuts on the clothing, and to any soil on the knees or the shoes. Glass particles would not have existed as debris until a piece of glass was shattered, so they alone are markers of the burglary. The glass would not have had broken edges until it was shattered, so any firmly adhering fibers, hairs, or blood would also be markers. These individual types of materials are an inherent part of the incident, thus relevant to the major case question; i.e., who broke the window and entered the building? Provided they are foreign debris, these materials are significant in establishing links between the perpetrator and the event.

In another example, if a man entered a woman's bedroom and raped her digitally, there would be no transfer of semen or saliva. The only type of material that may characterize the event would be any traces of vaginal secretion on the intruder's finger. Suppose the assailant washed his hands, destroying the only marker evidence. In this case, the remaining individual types of material would be far less useful, because they do not characterize the event. However, it may be possible to establish contact, if not rape, by evaluating the sets of debris on the clothing of each individual — i.e., the lint and the dust — and whether there is evidence of transfer from one person to the other and vice versa. This will be discussed more extensively later in this chapter and is illustrated in Case Examples 8.9 and 8.11.

8.4 Component vs. Non-Component Debris

Fabrics accumulate lint and debris with wear and use. The lint includes fibers and fiber fragments from the constituent materials of the fabric, referred to as component debris (Cwiklik, 1999). This component debris can be transferred to other items and is the experimental subject of shedding studies (Coxon et al., 1992). When transferred, it becomes the non-component debris of the recipient item. Non-component particles and fibers are those produced from something other than the item itself and are not an intrinsic (component) part of the item. When two items are in contact, both component and non-component debris are transferred. When several items of clothing are being worn at the same time and are in contact with one another, debris from one transfers to the others, producing a subset of debris that includes a high proportion of component fibers and defines that set of clothes (Figure 8.1).

8.5 Transfers of Debris

Traces and debris are deposited in three ways: transfer by contact, motion (smears and so forth), and projection. Contact and motion include touch and impact. Projection can be passive, such as a flour spill or falling ceiling debris, or dynamic, such as glass shattering, dust being kicked up, or sand thrown in someone's face and landing on the shirt below. It can result from suspended material, such as aerosol from paint spray, or a haze of fine fibers, such as from a cutting room in a garment factory. The suggested terminology for stains and deposits (see Section 4.10) can also be applied to traces and debris. It is useful to explore how several different types of traces tend to appear on clothing items and how they can be evaluated for further work. Several types of stains and deposits are discussed here as well.

Figure 8.1 Accumulated debris on black angora wool sweater, including animal hairs and blue-green fibers attributable to a blanket worn around the sweater.

8.5.1 Transfers of Individual Types of Material

8.5.1.1 Paint

Paint traces that appear in debris include paint that is flaked off already-peeling surfaces; has chipped off, either from preexisting damage or damage from the incident; has been deposited as an aerosol from paint spray (resulting in microscopic spheres, Figure 8.2); is in powder form from sanding or sandblasting; or has peeled off another surface. Weathered paint may form a powder upon impact or even from lighter contact (Figure 8.3). Any of these types of particles can be characteristic of a site or of the debris on a person. They can each be an inherent part of an event, such as the microscopic particles of paint spray described in Case Example 4.9, and can provide clues to the scene, like the flakes of peeling paint in Case Example 8.3. The particles that accompany paint traces may also provide clues, such as sand from sandblasting of painted surfaces, abrasives from sanding, metal or plastic from vehicle impact, and so forth. Wet and sticky paint transfers and paint smears deposited by impact are discussed in Chapter 4.

When a pedestrian has been struck by an unknown vehicle, paint on the clothing can provide investigative information about the type of vehicle involved. The examiner may encounter several types and colors of paint chips. Many of them may be part of road debris from that particular site and are unrelated to the incident under investigation. Such paint chips should be evaluated with respect to any smears of impacted paint. The impacted

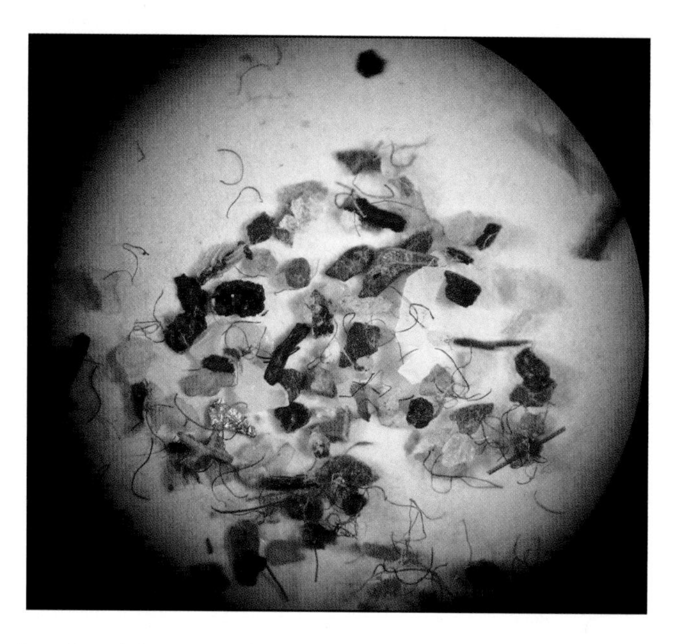

Figure 8.2 Sieved pocket debris includes several tiny chips of blue paint.

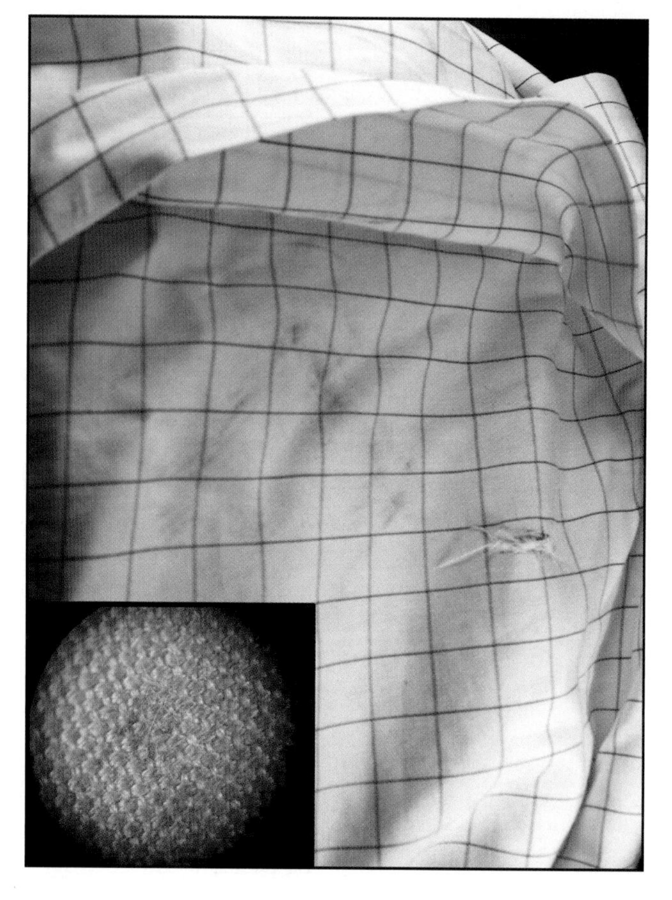

Figure 8.3 (Main Figure) Reddish-tan smears on the back of a shirt. (Inset) Small, dark-red particles of weathered paint observed at 40× magnification.

paint smears may be produced by the surface layer only and provide no information about primer or undercoats. However, paint smears are usually more significant than paint chips, because they provide information not only about the potential source of the paint, but also about what happened; that is, activity-level information (discussed in Section 4.2.4; Cook et al., 1998). Paint chips with the same type of surface coat can then be used to narrow down the vehicle in question, by examination of primers and undercoats, the sequence of which may be distinctive in a repaint. When the top layer is a clear coat, the examiner might find traces of the paint layer beneath it.

Debris, damage, and defects on the paint flakes and paint chips can also be significant, probably more so in architectural paints (i.e., paints used in houses and other buildings). If a victim is found in a location other than the crime scene, a foreign paint chip or flake can provide clues to the location of the crime scene if the debris is distinctive enough to narrow it down. Dust and debris from industrial or agricultural sites may settle onto the paint. So may detritus from trees and other plants. On occasion, mold growth is encountered and, for vehicle paint, may provide clues to the geographical origin of a vehicle.

CASE EXAMPLE 8.3

Paint flakes link a body to a murder scene; bindings provide link to scene and other cases; clues from individual types of materials; stronger links to scene via debris; link to suspect via personal recognizance confirmed by hair transfers

When a teenage girl who walked home from school every day did not arrive home one day at the expected time, police were called. They canvassed the neighborhood to no avail, and several days later, someone found the girl's body in an alley. She was wearing an Afro-style wig. The clothing, wig, and autopsy samples were searched in the laboratory on an urgent basis. Several large flakes of peeling paint were found entangled in the wig. Police searched a nearby vacant house and found peeling paint on the ceilings of the downstairs rooms. The scientist went to the scene and collected reference samples of paint and debris. The paint flakes in the wig corresponded with three colors of peeling paint from the house, all from the same room.

In the same room, police found a pile of pantyhose legs and pairs of pantyhose, each with one leg cut off. The victim had been bound with tightly knotted nylon pantyhose around her wrists and ankles. One of the bindings was a pair of pantyhose with one leg cut off and the other binding was a cut-off pantyhose leg. The two pieces of pantyhose were compared using physical matching. The separate (cut-off) leg did not match with the panty portion used to bind the victim, but it did match one of the pantyhose found at the vacant house. However, no corresponding leg piece was found at the house for the partial pair of pantyhose used to bind the victim.

The victim had been strangled, so there was little bloodshed, and because she was wearing a wig, she would not have shed many hairs. Follow-up laboratory examinations were conducted. Predominant types of paint, animal hairs, and fibers from the victim's clothing and wig were found to correspond with debris from the suspected crime scene. Several wig fibers from the scene corresponded with the fibers of her wig. A foreign pubic hair was found in pubic combings, although no semen was found in vaginal or rectal swabs taken at autopsy. Several transients were known to frequent the house, and police had still not narrowed down a suspect for the murder.

There was one partial pair of pantyhose from the scene that did not have a corresponding pantyhose leg to complete it. The laboratory had examined several recent cases in which women reported being tied up and then raped. In each of these cases, the bindings consisted of knotted pantyhose and pantyhose legs. The bindings from each of the cases were reexamined, and the missing leg from the homicide victim's pantyhose bindings was found among the evidence from one of those prior rape cases. Two of the rape victims were able to describe their assailant. Police apprehended one of the transients who frequented the vacant house. A reference sample of his pubic hair was obtained and found to correspond with the foreign hair in the homicide victim's pubic combings.

This case dates to a time before DNA testing was available. Had it occurred more recently, the rape cases, but not the homicide, would probably have been linked by DNA testing. Because no semen was found on the homicide victim, trace evidence and the pantyhose matches would still be the most significant evidence, together with any DNA that might be obtained from the foreign pubic hair, and the paint chips would still have provided the clue that led police to the abandoned house.

8.5.1.2 Glass

Glass has many uses in everyday life. It breaks easily and ejects very small fragments in different directions that may be retained by garments. Glass on clothing may be found in cases of homicide, burglary, hit-and-run vehicle accidents, and assault. The glass on the clothing can be compared to samples from its possible source to determine possible common origin. Glass fragments found on the clothing of a suspect can be compared with glass taken from a broken window at a burglary scene. Pieces of glass on a victim's clothing in a hit-and-run can be compared with glass from a broken vehicle windshield.

When loose glass shards are found on clothing, the shards are more likely to be directly deposited from breaking glass than transferred from debris. The usual poor retention of freshly broken glass militates against significant secondary transfer to clothing unless the shards are very small. Glass found on the soles and heels of footwear is the exception, because it can be scattered on the ground while shattering, then transfer from the ground to the shoe soles and retained in the treads.

The glass fragments embedded in the soles and heels when an individual walks over broken glass can be compared to a particular window, but rarely can it be linked to the act of breaking the window. Because glass from a broken window is often found on pavement below, a number of people may walk over shards of freshly broken glass. Fewer people may walk over broken window glass found on soil in a less accessible spot, and the traffic in such an area can be evaluated by footwear impressions left in the soil. Fresh impressions can be compared with shoes in evidence. Moreover, glass deposits on shoe soles can persist for some time, especially if they are retained in soil crusts or tar (Figure 8.4). The deposits on a pair of shoes can be evaluated with respect to other debris. Glass shards are most likely recent if found on the surface of soil or tarry material, especially if found on top of other debris.

Figure 8.4 Debris embedded in tarry deposit on shoe sole, including glass of non-recent breakage, small stones, plant parts, and yellow road or curb paint.

Glass can also be found in pockets and pants cuffs, where it can be protected and carried for quite some time. Tempered glass seems especially likely to be found in pockets. When tempered glass shatters, it produces rough cubes instead of shards and tends to fly off sideways as well as directly toward the person standing in front of it. In the personal experience of one of the authors, who cannot always resist touching cracked, but not yet shattered, tempered glass windows, the sideways-directed glass sometimes goes directly into side pockets. After it is carried around for some time, the microscopic jagged edges on the sides of the "cubes" tend to pick up particulate and fibrous debris. It can thus be recognized as debris rather than a fresh deposit.

When clothing from vehicle–pedestrian collisions is being examined for information about the vehicle that struck the person, not only paint but also glass is of interest, because the clothing may exhibit broken windshield, headlamp, side mirror, or even side window tempered glass. Although glass may not provide clues to the vehicle, it can later be compared with vehicle reference samples. The scientist should be able to distinguish freshly broken glass from debris glass. Only the former has the potential to be evidence of recent impact.

Glass from many sources — broken bottles, broken windows, broken decorative glass, broken picture glass, etc. — is a common component of debris in city streets (McQuillian, 1992) and is subjected to further breakage from footwear and wheels, which break off the sharp fracture edges and eventually produce somewhat rounded particles of glass. When rounded glass is found, it is almost certainly from non-recent breakage. At this stage, the glass, and the other debris it is now a part of, can be readily deposited and retained on clothing, then transferred to other clothing in secondary transfer of debris.

A tool used to break glass may retain crushed glass where it struck. There may be visible shards or a white powder comprising microscopic glass shards. Crushed glass can transfer to items of clothing and may provide a link between the tool used to break the glass and the person who wielded the tool.

Occupational exposure to glass shards may produce recently deposited glass shards from many sources. The retention rate remains low, but for the clothing of a glazier or

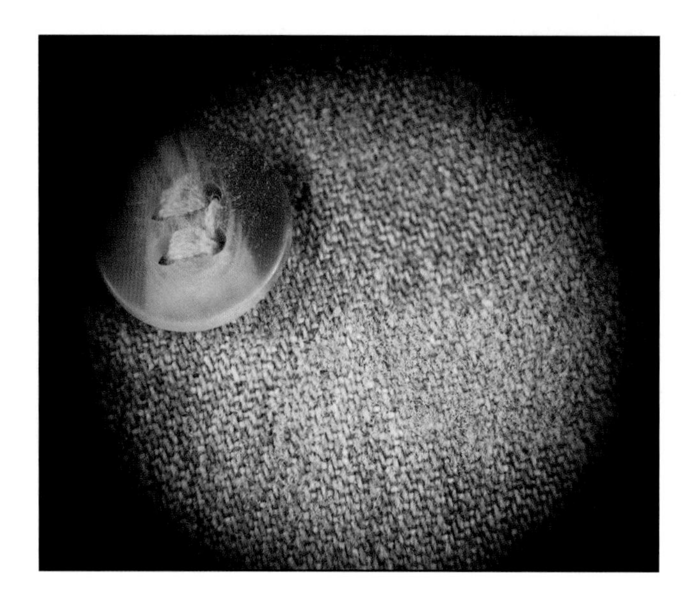

Figure 8.5 The cuff of a sleeve rubbed against volcanic ash (gray powdery deposit). The button exhibits scratches and the button thread fibers are partially cut.

a hardware store employee who frequently cuts glass, the number of shards deposited is high enough to result in a large number retained despite the much larger number lost. The clothing of a career criminal who specializes in break-ins may also exhibit a high number of glass shards.

When clothing is exposed to glass fragments, it often exhibits tiny, sharp, intermittent glass cuts as well. Even in the absence of glass particles, glass cuts themselves can indicate exposure to glass, because the cuts are highly characteristic (Griffin, 2007).

Finally, some glass deposits and glass damage to clothing are produced by volcanic glass. Some types of lava deposits comprise volcanic glass (obsidian) that acts like manufactured glass in producing damage to clothing and footwear. When chipped, it produces shards of glass. Examiners in regions with areas of volcanic flow, whether ancient or contemporary, should obtain reference samples of volcanic glasses for use in casework.

Volcanic ash includes numerous shards of glassy silicates, and glassy fringes on mineral grains and crystals (Cwiklik, 1981). The larger particles in a fresh deposit produce abrasions consisting of microscopic glass cuts on clothing and footwear (Figure 8.5).

8.5.1.3 Hair

Human hair can be linked with an individual to a high degree of specificity if a nuclear DNA profile can be developed from the root tissue. Most hairs encountered in casework are shed hairs with little remaining nuclear DNA, thus limiting the potential of this method. However, microscopic comparison and mitochondrial DNA testing, albeit less specific, can significantly narrow down potential sources, especially when used in tandem (Bisbing, 2007).

With the advent of highly specific DNA testing, microscopic hair comparison, which is far less specific as a link with a particular individual, seldom stands alone. Apart from the question of specificity is the question of reliability, called into question by prominent cases in which individuals were convicted of crimes based on microscopic hair comparisons that provided erroneous links. As a consequence, many laboratories have abandoned microscopic hair comparison altogether and have deemphasized the role of hairs as evidence. This deprives the criminal justice system of a valuable associative tool.

It is beyond the scope of a book on clothing examination to discuss the questions surrounding microscopic hair examination, including the new sources of error introduced when a DNA-only approach is used (Cwiklik, 2001, 2003; Bisbing, 2007). The bottom line for the clothing examiner is that hair is valuable evidence that should be routinely collected (Taupin, 2004).

Because of the potential of a single hair to provide a link with a specific individual, all the hairs on an item should be collected, rather than just a sampling of the hairs. An exception is a search for hairs from a victim on items from a suspect. Hairs on a suspect are significant only if the suspect is linked with the incident under investigation. Thus, an initial search of a suspect's clothing might focus only on those hairs that have the potential to be from the victim. If any are found, the remaining hairs should be collected as well, because they may be used in addressing the nature of any association between suspect and victim. For example, the analyst might address whether hairs attributable to a victim were deposited on the suspect's clothing via direct transfer or through an intermediary source. It may be possible to answer this question by studying transfers of debris, including hairs. Establishing transfer through an intermediary source is further discussed later in this chapter and is illustrated in Case Example 8.1.

Other information from hair microscopy includes somatic (body area) origin, growth stage of the hair root, length of the hair, whether the hair was recently cut, cosmetic treatment, damage, and adhering materials (Ogle and Fox, 1998; Robertson, 1999). Whether a hair was smashed by impact with a weapon or cut by an impact that shattered a windshield can be determined by the type of damage to the hair shaft. Hair evidence can be used in reconstruction of events by providing time line information from putrefaction or cosmetic treatment and activity-level information from type of damage. This is illustrated in Case Example 8.4. Hair that has been pulled out in clumps can be evidence of a struggle or assault. Hair that has been around awhile as part of debris can often be distinguished from fresher deposits by examination of damage and adhering small particles. Such a finding would argue against the hair being deposited by direct transfer during a contemporaneous incident.

Human and animal hair — from all parts of the body — is shed and replenished daily. A hair shaft emerges from a "root" enmeshed in a subcutaneous root sac. The root is living tissue, but the shaft is not. For a hair to be dislodged during the growth phase, it must be plucked out. Single hairs can be plucked with relatively little force by getting caught in a comb, on a necklace or lanyard, and by pulling. For hair to be pulled out in a clump, considerably more force is required. Because of the force required to remove them, clumps of hair found on clothing of a living person usually suggest assault or other trauma.

There may be reasons other than forcible pulling for clumps of hair to be deposited on clothing. This can be determined by a hair examiner upon microscopic inspection of the roots. During the clothing examination itself, all hair clumps should be collected and packaged separately from individually deposited hairs. Packaging the hairs in plastic bags allows the hair examiner to perform a rapid assessment of the growth stage and condition of the roots. A good preliminary assessment can be made using a stereomicroscope. Traumatic removal is suggested if most of the hairs exhibit roots that are in the growth (anagen) phase. If a person is undergoing chemotherapy and the roots in a clump of hair are atrophied, no extraordinary force need be involved. In homicide cases, once a corpse has begun to decay, the anagen roots and eventually the entire scalp putrefy and loosen. Clumps of hair found on the clothing when a loosened scalp is dislodged by scavengers should not be confused with clumps of hair pulled out with force. The former can be distinguished by the putrefaction of the hair roots (Petraco and Fraas, 1988; Linch and Prahlow, 2001).

Once the growth phase of a hair has ended, the root begins to withdraw from the sac and eventually forms a bulb that is pushed out, resulting in shed hair. Most shed hairs, unless very short, remain with the actively growing hairs until mechanically dislodged. Hairs are dislodged frequently and by a number of activities. The movement of unbound hair dislodges loose hairs, which fall onto upper clothing, furniture, bedding, and so on. Very short loose hairs are swept down the drain when the scalp is rubbed by washing. Hairs are dislodged by touching, transferring to the hands, clothing, vehicle seats, and via the hands to a person's pockets. Scalp hairs are found in the pockets even of individuals with very short hair. Pubic hairs are often found in underwear and on bathroom and bedroom floors with other body hairs and scalp hairs. Loose hairs are transferred by direct contact onto headgear, towels, bedding, vehicle seat head rests, and the clothing of other persons during close contact, and are then transferred from those items to other clothing.

Animal hairs are shed in sleeping areas, against rubbing posts, fences, and trees, and directly or indirectly to the clothing of household members or handlers. The clothing of individuals with animals in their households or who are in contact with animals in their daily work usually exhibits a collection of animal hair. A transfer of animal hair to the clothing of another person can provide the basis for association, especially if other debris has transferred as well (Suzanski, 1989). DNA testing has allowed associations between a deposit of animal hair and a particular animal, a goal that has not been attained for individual animal hairs using microscopic hair comparison. Although the degree of specificity possible with human DNA is not yet equaled in testing animal DNA, strong links can nonetheless be provided. Microscopy can be useful in studying a hair deposit that includes a range of characteristics if the range is reflected in a reference sample from the candidate animal. The clothing examiner should ensure that not only the larger guard hairs are collected, but also finer fur hairs. Fur hairs from fur garments are usually sampled and examined as fibers.

The accumulation of loose hairs on the upper back of a shirt or sweater can be used as a secondary control; i.e., presumed, but not known, to originate from the wearer. A secondary control may be used in missing person cases where kidnapping is suspected and evidence of the missing person is sought on a suspect or in a vehicle. Hairs from the upper back of a garment can also be used as evidence of the owner or usual wearer of the garment (Figure 8.6). A finding of several hairs in frequently used pockets can provide lesser evidence of the wearer, especially in the context of other characteristic debris.

Once transferred, hairs become part of the debris on an object or person and are subject to subsequent transfer to other objects. Hairs found on a pair of undershorts may have transferred from the bathroom floor. Secondary transfer is frequent (Grieve and Biermann, 1997; Suzanski, 1989; Taupin, 1996). Debris on the clothing of an individual typically includes hairs not only from that person, but also from members of the household and other close contacts. When the clothing is in contact with another person's clothing, a transfer of debris includes a sampling of the entire set of hairs on the clothing, foreign hairs as well as "native." This is illustrated in Case Example 8.1. Also illustrated in the case

Figure 8.6 Deposits of hair on the inside upper back of a sweater, suitable for use as a secondary reference sample.

example is a context-based method of using the entire set of debris to evaluate whether the transfer was primary or through an intermediary source (Cwiklik, 1999, 2001). In the latter case, hair is treated less as an individual type of material than as a member of a set of debris.

Pubic hairs may often be transferred during sexual contact, but the persistence rate is low, probably because the transferred hairs are readily dislodged during urination and hygiene. Despite the relatively low rate of persistence in pubic combings, pubic hairs can be highly significant when found. Foreign pubic hairs may persist in underpants or undershorts even if absent in combings. Because the pubic area is usually protected by clothing, pubic hair is not part of general debris except in areas where people disrobe or sleep. The sources of secondary transfer are usually limited to those floors and bedding and to any foreign hairs already in the pubic hair of a sexual partner.

If a complainant reports being raped by several individuals, secondary transfer of hair can provide information about sequence of partners if pubic combings were obtained from all the individuals involved and if pubic hairs are found on underpants and undershorts. Hairs from the first assailant transfer to the victim's pubic hairs and underclothing, then to the second assailant, and so on. Testing of commingled hairs avoids the problem of commingled body fluids, because each hair retains its physical integrity (Cwiklik and Gleim, 2009). If nuclear DNA testing on such hairs is successful, no interpretation of mixtures is necessary. When insufficient DNA is available, microscopic comparison alone can eliminate individuals, thus narrowing down the possible donors. Microscopic comparison combined with mitochondrial DNA results can often provide strong links with a particular individual.

CASE EXAMPLE 8.4*

Cut hairs used to reconstruct cause of death

The badly decomposed body of a woman was found by hunters in a wooded area. No bullets were found, nor were any of the bones damaged by bullet or knife wounds. The hyoid bone of the neck, often broken in strangulation, was intact. The cause of death was not determined at autopsy. When the clothing was examined in the laboratory, the examiner noted a number of hair fragments — probably the victim's shoulder-length hair — near her neck. The hairs were roughly cut. The hair fragments were correlated with her own hair, found as a loosened scalp. This suggested that her throat had been cut and some of her hairs were cut at the same time.

Drag marks on the back of the clothing and backs of the shoes, together with the absence of mud and dirt on the shoe soles, suggested that the victim was killed elsewhere and her body transported to the wooded area, where it was dragged to the grave site.

* Information provided by Lynn D. McIntyre, Washington State Patrol Crime Laboratory, Seattle.

8.5.1.4 Fibers

As the constituent material of clothing, fibers are nearly as ubiquitous as debris. Debris found on clothing includes fibers from other clothing worn by the same person as well as fibers from furniture, blankets, fabric toys, and clothing from other people in the same household. Fibers from fabric sources in other environments of an individual, such as a workplace or a close relative's house, are also reflected in debris on the clothing. Transfer of fibers could be expected to occur in the manner proposed by Grieve and Biermann (1997) for experimental transfer of clothing fibers; i.e., primarily through intermediary articles of clothing and only secondarily by direct transfer. One would also expect articles such as bedding and furniture to serve as intermediaries for transfer. Intermediary transfer is further discussed in Section 8.5.2.

If only certain types of fibers are of interest — for example, if clothing from a suspect is being compared with possible source fabrics from a victim or a site — fibers can be collected using forceps and placed into small zip-lock style plastic bags. This allows the trace evidence examiner to perform a quick screen to determine whether further examinations will be needed. Additional fibers can be sampled on sticky-notes such as Post-it® notes or on clear wide tape. The sticky-notes can be labeled directly then placed into plastic bags for rapid screening, and tape can be labeled and affixed to clear plastic sheets such as document protectors.

If fibers are being collected from a victim's clothing, usually the entire range of fibers is of interest, because it is not known whether the suspect is the true perpetrator. Of those collected, subsamples of certain fibers may later be selected for further attention. However, the initial sampling should be comprehensive. Collection with forceps is recommended, even for very small fibers.

When a transfer of fibers does not initially seem to address the case questions, a just-in-case sampling should be performed. Debris from the garment or selected areas of the garment can be sampled using sticky-notes or clear tape as described above. Sticky-notes have the advantage of weaker adhesive, which permits easier removal of particles and fibers, and the disadvantage of being opaque. Clear tape has the advantage of permitting a view of both sides of the adhering particles and fibers. A solvent is usually required to remove fibers and particles from clear sticky-tape unless water-soluble tape is used (Chable et al., 1994).

Fibers collected from other sources, such as the edge of a knife blade or an exposed edge of a broken window, may be submitted for comparison with clothing items. Fibers are not unique, and depending on case circumstances, intermediary transfer may be possible. Therefore, the clothing should be examined not only for correspondence of the fibers, but also for any damage to the clothing that might have been produced by the event resulting in fiber deposits, whether it be glass cuts, abrasion, or knife cuts to the fabric.

8.5.1.5 *Gunpowder Particles*

Gunpowder particles from ammunition are usually encountered as a result of firearms discharge. If a person sustained a bullet wound, a pattern formed by burnt and unburnt gunpowder residue around the bullet entry hole can provide information about firing distance. However, such a pattern is seldom seen if the firing distance exceeds three feet. The use of gunpowder particles in determinations of firing distance is discussed in Chapter 5. Gunpowder particles are found among general debris in regions where firearms are in wide use — in roadways in many rural and forested areas, in campsites, and in city streets and sidewalks. Debris from the clothing of hit-and-run victims often includes gunpowder particles even when no firearms were involved and the victim does not own or use a firearm. Unless gunpowder residue is observed as a pattern around a bullet hole, it usually has little value. Exceptions are finding such particles in restricted areas, in enclosed spaces such as vehicle interiors, or when reference debris from the site does not include gunpowder particles. In those cases, limited case-specific conclusions may be drawn if reference samples of debris do not include gunpowder particles.

Specific types of gunpowder particles that are part of a set of debris may figure in associations based upon debris.

CASE EXAMPLE 8.5

Loose gunpowder particles in clothing debris

A bullet hole was found in the knit nylon shirt of a man who was shot in a car, but no pattern of gunshot residue was observed. The fabric retained little debris of any type. However, numerous burnt and unburnt gunpowder particles were found in debris that fell off the shirt while the clothing was being dried prior to packaging. This indicated that the gunpowder particles were deposited but not retained and suggested that the firing distance was within approximately three feet. However, it was neither a contact shot nor a near-contact shot. In those cases, embedded gunpowder particles would be expected from the force of such a shot; this was not observed.

8.5.1.6 *Soil and Sand*

Soil deposits are often found on items of clothing, as mud splashes, caked soil deposited during a struggle, caked and smeared when a body was dragged from one site to another, loosely smeared, ground into the knees, transferred to a pocket, or crusted with blood. Sand may be found crusted with blood or bound by a clay component. Loose sand is less often observed on clothing surfaces because it tends to fall off, but it can be found in pants cuffs, pockets, and rolled-up sleeves, and in the scalp and headgear. If a person is partly or completely submerged in a body of water, there may be sand even on the inside of clothing. Both soil and sand are found on shoe soles, often in the tread design elements of a patterned sole.

The color, texture, type, and amount of vegetation, size and size range of the mineral grains, pollen assemblages, and artifacts of human use are properties of soil that can be observed with a stereomicroscope in soil deposits on clothing. Human artifacts include traces of building materials such as glass, paint, bricks, concrete, plaster, and wood as well as industrial residues and agricultural additives. If several colors of soil or types of deposits are observed (whether loose, caked, mud-splashed, etc.), the item of clothing may reflect more than one environment. The environments may be several feet away from each other or, in areas of windblown soil that shifts and blows, many miles.

Ideally, the person who collects soil samples from the clothing should be the person who examines the soil. The appearance of the deposit may govern the technique used for sample collection and may even influence the method of comparison. For example, when soil is caked onto clothing or footwear, it may be deposited in layers. Each layer should be sampled separately. When soil is encountered as a light deposit, it may represent only the smaller fractions of the source soil; a light deposit on clothing may have a different appearance, even in color, than samples of a potential source soil, yet the finer fractions may correspond. The evidence sample and the soil reference sample should both be sieved and the corresponding fractions compared separately, because only the lighter fractions may be retained on the clothing. Soil splashed onto clothing from a mud puddle may consist primarily of the clay fraction; the very small clay particles may be suspended in water even when larger particles have settled to the bottom. The color and composition of clay can be useful in characterizing a soil sample even if it is only one component.

Caked deposits of soil and encrusted deposits of sand can be gently lifted with a spatula. Place onto a clean piece of paper for preliminary examination, or for storage, place into a cardboard or metal box with a tight lid. Take care that the sample container is undisturbed, so that layers remain intact. Otherwise, the layers should be immediately separated before the sample crumbles. Deposits of soil and sand found in pants cuffs or loose in pockets can be collected by letting the contents drop onto a piece of clean paper. Package any clumps of soil like a caked deposit. Loose deposits or light deposits of soil or sand on the surface of a clothing item can be gently scraped onto a piece of clean paper placed directly adjacent to the deposit. It is often helpful to tap the substrate fabric from the opposite side to dislodge a thin deposit. Avoid scraping fibers from the substrate fabric as much as possible. Small deposits can be collected using a needle probe with a bit of water-soluble adhesive on the tip and transferred directly to a microscope slide (Teetsov, 2002).

CASE EXAMPLE 8.6

Sand deposits used to narrow down where a body entered the water

The body of a man was found in a shallow Puget Sound cove and foul play was suspected. Investigators wanted to know where he had entered the water. Sand found on the clothing included not only beach sand from the cove where the body was found, but also pockets of sand from another site. The movement of objects in the water currents was sufficiently well known that the marine scientists consulted by police narrowed down the most likely places where the body could have entered the water, including areas where a body could have been thrown from a boat or ship. The sand from the clothing that was not from the beach where the body came to rest was compared with reference samples from the candidate locations. All but two of the candidate sites were eliminated, allowing police to better focus the investigation.

8.5.1.7 *Pollen, Spores, Wood, and Other Plant Parts*

Plant detritus that is found on clothing reflects the locations where the clothing — and usually the person wearing the clothing — has been. Plant materials commonly found on clothing include stickers and burrs, thorns and needles, trichomes, seeds, spores, and pollen. Also included are residues from mosses, lichens, algae, and fungal bodies. Plant stems, woody parts, barks, and leaves may also be found on clothing. Although few crime laboratories are staffed with specialists in botany (the study of plants), mycology (the study of fungi), or palynology (the study of pollen), a good microscopist with a good reference collection can learn to identify many plant materials. Full interpretation of findings requires knowledge of occurrence, distribution, expected companion deposits, and alternative sources of the plant parts in question. However, even a nonspecialist in these areas can perform useful preliminary examinations and can consult with a specialist as needed. Whether the specialist is in the same laboratory as the clothing examiner or is a consultant from outside the laboratory, the clothing examiner is responsible for recognizing and collecting the materials of interest and is in the best position to evaluate deposits with respect to other evidence on the clothing.

Palynology, reported in casework and taught in microscopy courses for at least 50 years (Horrocks et al., 1998), is an emerging forensic science discipline, with attention to transfer, persistence, and significance as associative evidence. Pollen is found dusted onto clothing from direct contact with pollen-bearing plants and is a component of mud and soil found on clothing. It can thus be used to link a suspect or victim with a scene, link a deceased person with a murder scene different from the grave site, and link persons with vehicles that may contain dust from the shoes or belongings of people who have been at pollen-bearing sites. Pollen deposited on clothing from direct contact with flowering plants is often yellow, orange, or reddish dust. A dusting of pollen can be transferred to a microscope slide with adhesive material on a needle probe, as described for soil deposits. Pollen that is part of mud or soil is usually isolated to be further analyzed. The reader is referred to references on palynology for pollen sample preparation. Because many types of pollen are airborne, it is important to avoid contamination from other pollen sources during collection, storage, and laboratory examination of clothing items.

Spores can be treated much like pollen, with the exception of suspected anthrax spores, which must be treated under protocols for biohazards. Anthrax is resident in the soil of certain regions, such as the southeastern United States, particularly in rural areas.

Wood deposits on clothing originate from trees, stumps, and logging debris (Cwiklik, 1999, Case Example 8) as well as from various stages of its use as an architectural material. Deposits include wood chips, sawdust, smashed wood (Case Example 4.11) and wood splinters, as well as wood bark and cork fragments. Wood is the raw material for most types of paper; the processed form can be distinguished from raw wood even if only a few wood fibers are available. Sampling protocols depend on the form of the deposit rather than the material. Wood dust can be sampled with an adhesive-tipped needle probe, splinters are collected with forceps, and sawdust is collected using methods described for soil of comparable size range and aggregation.

Other plant parts include burrs, stickers, leaves, seeds and seed husks, grasses, stems, parts of flowers and fruits, and fungal bodies. Plant parts on clothing are especially useful if they indicate a location different from the site where the clothing or the person wearing the clothing was found, because plants may reflect vegetation at the crime scene. If a body is found long after the disappearance of the person, the stage and extent of plant growth may indicate the season in which the body was left at a gravesite and sometimes the elapsed years. Like soil, vegetation on clothing may corroborate or cast doubt on an account of events at a particular site. For example, a woman may report being raped in a wooded area away from the grassy picnic area that she and the suspect both agree was their original destination. If the suspect claims not to have been in the wooded area yet both their clothing exhibits that type of vegetation, her account is supported and his is not. Vegetation may also be from agricultural or horticultural plantings or storage areas or from gardens or other plantings.

Finally, assemblages of types of vegetation, whether pollen or other plant parts, may be highly characteristic of restricted areas or types of sites. Interpretation of plant or pollen assemblages may require the expertise of a botanist, palynologist, field biologist, forester, or agricultural specialist.

8.5.1.8 Insects and Insect Parts

Insects and arachnids (spider family) are the often-uninvited companions of humans and animals in life and in death. Insect legs, wings, and droppings from household debris and outdoor sites may be found on clothing items. An entomologist may be able to glean information about geographical location from the assemblage of insect parts, and even a non-specialist can conduct comparisons of insect parts in sets of debris.

After death, the types of insects and the stages of their growth allow the entomologist to reach conclusions about time elapsed since death. Living insects are usually collected at autopsy, but dead insects found on the clothing of a homicide victim should also be preserved and shipped to a forensic entomologist for analysis. Mayflies are among the first insects to appear on a corpse, laying eggs in mucous membranes and open wounds, where their larvae, maggots, can feed. The evidence on clothing items of maggot activity consists of masses of flat white sacs.

CASE EXAMPLE 8.7

Maggot sacs used in reconstruction of death

The skeletonized remains of a body were found in a wooded area, scattered by scavengers. The skull was found some distance from the rest of the skeleton at a later time, and the connecting neck bones were never found. An accumulation of maggot sacs was observed during the clothing examination on the collar of the victim's sweatshirt — not just at the throat, but all around. This suggested an open wound consistent with decapitation. This was confirmed several months later when the skull was found.

8.5.1.9 Cosmetics and Glitter

Cosmetics include a broad range of materials including facial foundation makeup, lipstick, eye shadow, eyeliner, mascara, blush, nail polish, and hair products. Tiny reflective geometric pieces of glitter are now included in the list and have been the subject of recent study (Blackledge, 2007). Foundation makeup, lipstick, blush, and less often mascara or eyeliner, may be encountered as smears on clothing, including the clothing of assailants and consensual sexual partners who may brush up against a woman's face. (This is illustrated in Case Example 4.5; also see Chapter 5.)

Cosmetics may be thrown at an assailant in the course of a struggle or be transferred from the hands of the person who is applying them. Oil stains on clothing may be from hair oil products. Deposits of apparent cosmetics may permit an association of clothing from one individual with that of another or of either individual with a crime scene. The clothing examiner's preliminary examination should include comparison of color and microscopic texture.

The sampling of stains and deposits such as smears of liquid foundation, lipstick, nail polish, and hair oils is treated in Chapter 4. In general, a stain should be excised or extracted for any further testing, the specific solvent depending on the type of material. The sampling of a deposit depends on the form it takes, whether caked, crusted, glued, or otherwise adhering to the substrate.

8.5.1.10 *Foam Rubber and Plastics*

Foam rubber is found in upholstered furniture pillows, car seats, padding for sleeping bags, carpet padding, the under-the-surface layer of imitation-leather athletic shoes, and packaging material, and, as Styrofoam®, in food containers and microscope slide mailers. As these items break and degrade, particles of foam rubber and Styrofoam® become part of debris. Some products, such as pillows, use foam rubber already broken up into small chunks. Such particles are often found on clothing items, ranging in size from nearly microscopic to several centimeters long, and can be recognized by a honeycombed or flattened honeycomb appearance. Foam insulation residues may appear on the clothing of construction or home remodeling workers. The great variety of foam rubber types and the high transfer and persistence of such particles make it a forensically significant material (Wiggins et al., 2002). Anecdotal evidence can be provided by anyone who has opened a package containing Styrofoam® peanuts, or a pillow containing foam rubber chunks, and tried to clean up afterwards. Larger foam rubber particles are collected using forceps. Smaller particles are usually encountered in dust and debris (Figure 8.7).

Plastic is nearly ubiquitous in modern life, and particles of worn, torn, and chipped plastic are found in debris. Plastics are also a part of composite materials such as paint,

Figure 8.7 Foam rubber exposed by wear abrasion on the upper edge of an athletic shoe.

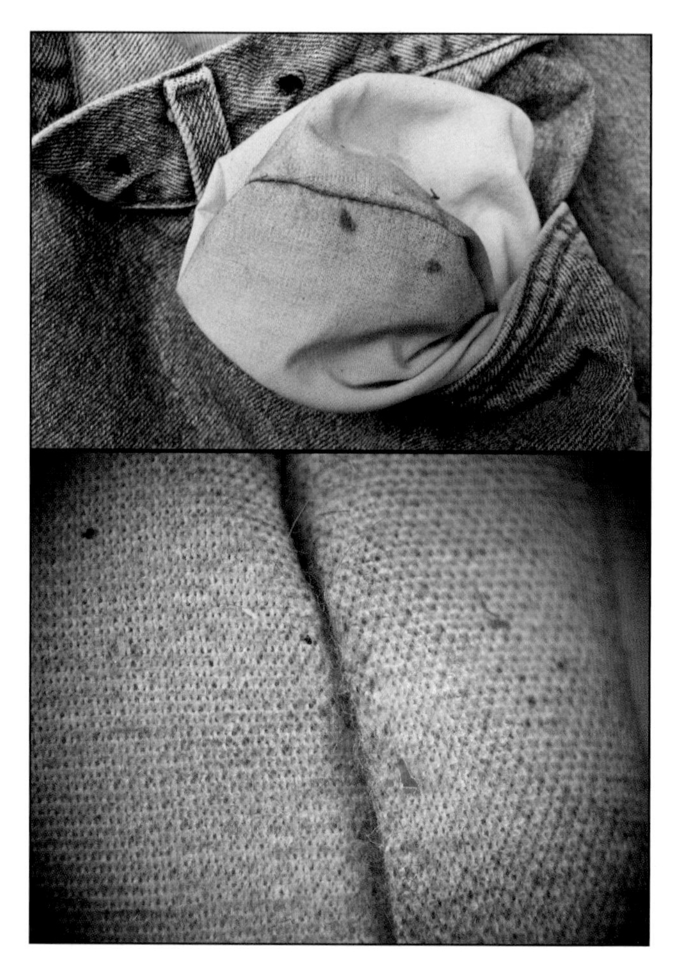

Figure 8.8 (Top) Pocket debris, including a blue plastic particle at the left of the photo. (Bottom) Particle of blue plastic at 16× magnification.

where it dries as a polymer, or fiberglass, where glass fibers are embedded in a plastic resin. Plastic particles in debris may reflect industrial processes such as spraying for heat deposition, where small plastic pellets are sprayed onto a heated metal surface or poured into a heated mold. This method is sometimes used in painting automotive panels. Tiny spheres may be observed from spraying paint or plastics. Droplets of plastic resin are used to fuse fiberglass insulation, and fibers in non-woven fabric rags are heat-fused in spots to hold the fibers together.

The very profusion of plastic particles may contribute to their often being overlooked. After recognizing the material as a plastic or plastic-containing material, the examiner must decide on a next step. If particles of similar appearance are found on an item from a potentially significant source, the particles may be segregated from general debris and submitted for microscopic and chemical testing and comparison (Figure 8.8).

However, if no comparison sample is available, it may be difficult to decide whether to pursue further analysis of plastic particles for investigative purposes. We suggest sorting by physical appearance. Spheres, droplets, chips, flakes, machined and abraded fragments have all been subjected to different mechanisms and may thus have different sources. Chips, flakes, and abraded fragments are probably wear or damage from larger items. Spheres and droplets reflect production processes. Decisions to perform further testing should flow from the case circumstances. If an unidentified body has been found, police and medical investigators may be interested in clues to the work and environments of the deceased. Links to a perpetrator are of interest in unsolved homicides. In a series of crimes, information about a transport vehicle may provide leads to the perpetrator. Consultation with experts outside the crime laboratory may be of value in determining the types of information that can be gleaned from plastics evidence in a specific case.

CASE EXAMPLE 8.8

Foam rubber, sand, and acid-damaged materials in debris linking a homicide victim to the trunk of a car

The body of a young woman was found in the wooded part of a marina. Sand found on her clothing was accompanied by bits of cardboard and pieces of foam rubber, some of it damaged by strong acid, with identifiable sulfuric acid residues. The sand was light tan, unlike the gray sand of the beach at the marina. Long blond hairs were found on her clothing, although her own hair was dark. The young blond-haired son of one of the marina owners was eventually suspected of the homicide. The trunk of his car was filled with debris, including chunks of foam rubber similar to that on the victim's clothing, a leaking automotive battery in a cardboard box, tan sand from a construction project corresponding with the sand on the clothing, dark hair microscopically similar to a reference sample representing the victim, and a collection of fibers and particles that corresponded with other debris on the victim's clothing. A blue paint smear was observed on a concrete post at the marina entrance. The blue paint of the suspect's car exhibited a scrape on the fender, and the blue paint smear on the post corresponded with the fender paint. Apparent blood found in the vehicle trunk was degraded by the battery acid and did not provide further information, but the traces and debris established a link with the vehicle and the long blond hairs on the victim's clothing established a link with the suspect.

8.5.1.11 Lubricants from Condoms, Contraceptive Creams, and Related Materials

Lubricants from condoms, vaginal lubricants, and contraceptive creams and jellies may also be found on clothing items. Although often reported with vaginal and anal swabs (Blackledge, 2007), deposits on clothing items can provide links between persons and the conditions of sexual activity in cases of reported rape or serve as details corroborating or casting doubt on an account of what transpired. Lubricants on clothing are found in trace amounts as direct deposits or as co-deposits with vaginal secretions. The deposits are often absorbed into the fabric and are difficult to find. If the deposit is concentrated, it may be observed as an oily stain; i.e., slightly translucent when examined with oblique illumination. A variable light source or ultraviolet light may be useful in locating such deposits if the fabric is fluorescent and if the deposit is not (R.D. Blackledge, personal communication). Co-deposits may serve as locators. For example, if a rape victim says that a perpetrator washed his hands, a towel on the bathroom floor may bear deposits of lubricants co-deposited with other material from the perpetrator's hands. If suspected but not observed, lubricants may be detected in extracts of a target area of the clothing such as

Figure 8.9 Black particles from smudge of charred wood. Note elongated fractured particles.

the crotch of underpants. Research is needed on methods for locating deposits of condom lubricants on fabric items, including clothing or bedding.

8.5.1.12 Soot and Other Black Smudges

Black smudges on clothing are usually either greasy or particulate. The color of grease deposits is usually from particulate contaminants in the grease, many of which are soot. Soot is essentially carbon dust. Particulate deposits of soot arise from anything that would produce carbon dust or from carbon-dust-based products such as charcoal, artists' or potters' pigment, or fingerprint dusting powder. Soot is produced by fires and high heat, burning candles, firearms discharge, smoke from automotive exhaust, and diesel exhaust (Figures 8.9 and 8.10). Soot residues from burning plastic-coated wiring are described in Case Example 8.2 (Figure 8.11). The individual carbon particles are beyond the resolution of the light microscope, but particle agglomerates can be recognized and associated chemicals tested. The latter can further identify the specific type of soot residue.

Black smudges other than carbon dust can arise from graphite lubricant, pencil marks, black mold, black eye shadow, and rubber dust, especially from tires. Tire dust can be transferred from roadways as well as from direct contact and may be important in reconstruction of vehicle–pedestrian accidents (Figure 8.12).

8.5.1.13 Beads and Spheres from Welding, Soldering, Burning, and Incineration

Welding and soldering both use heat to join metal. Solder is a lead-based metal usually used to secure a wire to a contact by joining the two metals with a bead of molten solder that hardens on cooling. Debris from soldering includes beads of the low-melting solder and thin tailings from its application. Welding is the high-temperature fusion of metal to

Figure 8.10 Black particles from smudge of diesel soot. Note fine particles coating individual fibers.

Figure 8.11 Black particles from smudge of burned insulated wires. Note wide size range.

Figure 8.12 Black particles from rubbing against a tire. The curled particles are often seen in rubber.

produce a metal seam. A metal such as aluminum alloy or steel is heated sufficiently that tiny molten beads of metal are produced. Hardened metal beads may be found on clothing, either loose or embedded in the fabric, often with accompanying thermal damage.

Debris from burning or incineration includes spheres and spheroids produced by molten minerals that form glasses or mineral melts and then harden upon cooling. Glassy beads may be from industrial flyash (McCrone, 1979) or from commercial incinerators. White spheres of calcium carbonate and hollow brown and black spheres appear on pyrolysis residues when cellulosic and other carbon-based materials are burned (Figure 8.13; Cwiklik and Dean, 2008). Brown and black hollow spheres are also encountered in gunshot residue from black powder weapons and as residue from pyrotechnics (Kosanke et al., 2006).

Metal spheres and other particles can be tested and compared with potential sources using elemental analysis. Black and brown spheres can be tested for organic content using FTIR or GC-MS (Fourier transform infrared spectrometry or gas chromatography–mass spectrometry). Glassy spheres and beads from mineral melt can be examined using polarized light microscopy then compared using elemental analysis and, depending on the composition, organic analysis as well.

8.5.1.14 Materials from Evidence Packaging

Debris from materials used in evidence packaging is often observed on items of clothing. The examiner should recognize them. A reference collection should include samples of paper sacks and butcher paper, Styrofoam, evidence tape, and other tape and plastic, including fragments of the beaded closure of zip-lock style plastic bags. The examiner should also recognize stray ink marks that may be from pens used to mark and initial the items and can use the initials on the clothing item and its packaging as reference points (Figures 8.14 and 8.15).

Figure 8.13 White spheres ($CaCO_2$) and hollow black and brown spheres in sieved ashes from T-shirt burn.

8.5.2 Transfer via Direct or Indirect Contact

When foreign debris on one item corresponds with the component material of a potential source item, it is not always clear whether the correspondence is due to direct contact, contact with an alternative source, or secondary transfer from the original source. Materials are often transferred through an intermediary source that was the recipient of the original transfer (Suzanski, 1989; Lowrie and Jackson, 1994; Grieve and Biermann, 1997; Taupin, 1996). When fibers like the component fibers of a garment or other fabric source are found on a garment from another person, direct contact is supported when most

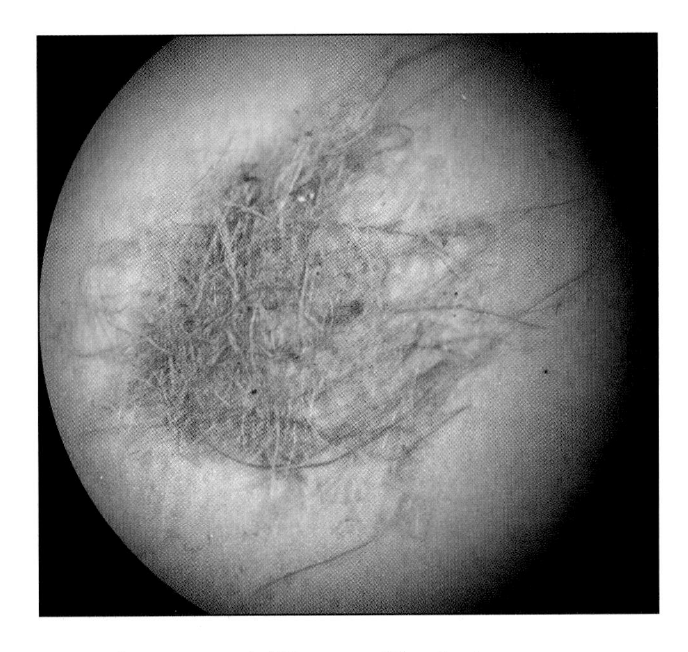

Figure 8.14 Fragments of a paper sack (25× magnification).

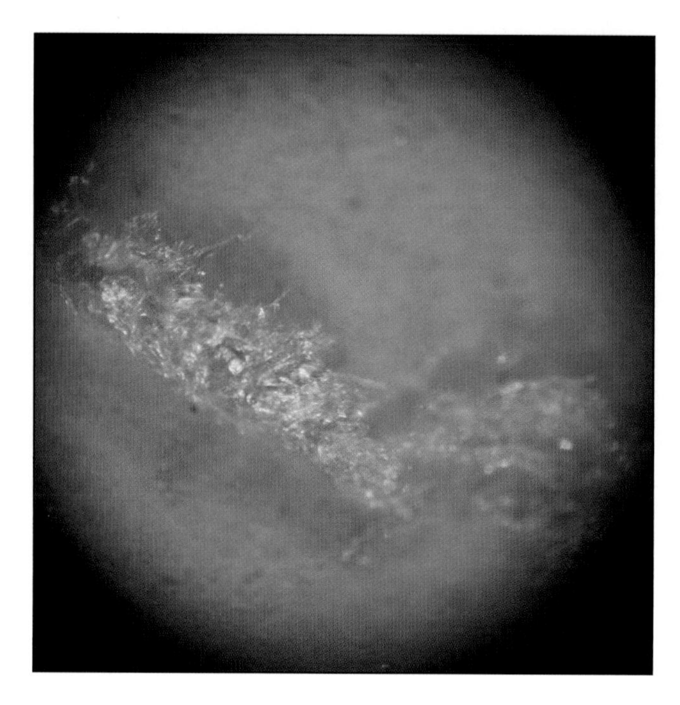

Figure 8.15 Fragment of a Styrofoam of similar "peanut" (40x magnification).

types of noncomponent debris on the first garment are also found on the second garment and few types of debris fail to correspond (Cwiklik, 1999). This degree of correspondence is usually observed only when the garments are minimally changed from the immediate post-contact condition. That is, the garments of neither individual were worn very long after the incident under investigation. This condition was met in Case Examples 8.1, 8.9, and 8.10.

In most cases, the garments of at least one of the individuals suspected of being in contact continue to be used after the contact incident, allowing transferred debris to be lost and additional debris to be deposited. It may then be possible to establish an association that includes a wider range of possibilities including both direct contact and contact through an intermediary source. Sometimes an association other than direct contact can be highly significant, and in other cases, a finding of indirect rather than direct contact may be exculpatory. This is discussed in Case Examples 8.11 and 8.9, respectively. The degree of correspondence of non-component debris influences the strength of an association, even if direct contact cannot be established.

Sometimes it is of interest to determine whether a garment was worn inside out, or in what sequence the layers of several garments were worn by an individual. The expectation is that surfaces that were in direct contact exhibit a higher number of component fibers from the adjacent garment and a close correspondence of non-component debris, whereas debris from the other surfaces exhibits a lower degree of correspondence. This must be evaluated with respect to the capacity for persistence and retention of each recipient fabric.

Finding potential sources for predominant non-component debris increases the strength of an association. When transfer through an intermediary source cannot be eliminated, the examiner should obtain reference samples of debris from logical intermediary sources. If direct contact with an intermediary source is indicated, direct contact with the source initially suspected of being in contact can be eliminated. Transfer via an intermediary source is illustrated in Case Example 8.1; the examiner concluded that hairs found on the clothing of a victim and attributable to the spouse of a suspect were deposited by secondary transfer from the suspect's sweatshirt.

CASE EXAMPLE 8.9

Direct transfer or transfer from an intermediary source?

Several students who shared an apartment in student housing had a birthday party for one of their group. The three students in the apartment next door, a married couple and a single woman, were their friends and were invited. The single woman left early to study for a test and went to bed early as well. She reported being digitally raped by an intruder she saw only by the light of her computer screen saver. She described the intruder as one of four men at the large party. Police collected her bedding and took her to the hospital, where swabs were taken and her clothing (a T-shirt) was collected. Police went to the residence of the man she described, a non-Caucasian student in his senior year and about to graduate. He had walked home alone, with no witnesses, and went to sleep without changing clothes. Police took him to the hospital as well, where medical swabs and clothing were collected. His hands were also swabbed for possible evidence of vaginal secretions. The sexual assault kits, clothing, and bedding were sent to the crime laboratory.

Laboratory test results yielded no foreign DNA from the vaginal or penile swabs, but the hand swabs from the young man yielded DNA from at least four individuals. The victim's DNA profile was included in the mixture from the hands, but the body fluid origin could not be identified, so it was not known whether the DNA could have been transferred during a handshake. Foreign hairs, including pubic hairs, were found on the bedclothes, but the hairs were of Caucasoid origin and unlike those of the suspect. The victim said that her last sexual intercourse had been six weeks earlier.

The suspect's clothing exhibited no hairs that could be attributed to the victim, but did exhibit four purple or maroon fibers of the same type, composition, and dye profile as the unusual fibers of the victim's bedspread comforter. The suspect was arrested, tried, and convicted. When the case was revisited two years later, the possibility of intermediary transfer was raised. No debris samples had been taken from the apartment where the party was held and where the young woman was a frequent visitor. Thus, it could not be established whether or not fibers from the comforter were a part of the debris in that household, especially on the furniture in the living room where the party was held. Nor were the clothing or bedding available for reexamination; they had been destroyed because of a clerical error. Neither the young woman's T-shirt nor the young man's clothing had been sampled for debris — only for target fibers.

Could a comparison of samples of debris from the clothing have indicated contact between the suspect and the victim or her bedding? Or would transfer of the purple fibers via intermediary contact be suggested, which could be exculpatory? A target search did not permit the case questions to be adequately addressed, and when the issue was revisited, it was too late.

8.5.3 Transfer, Persistence, and Detection

Three factors that affect findings of debris on clothing are transfer, persistence, and detection (Parybyk and Lokan, 1986; Robertson and Roux, 1999; Salter et al., 1984). Consider the following example. Two individuals — one wearing a nylon raincoat with a tight weave, the other wearing a wool sweater with traces of facial cosmetic powder transferred to the collar — engage in a tussle, and their clothing is studied afterwards for evidence of contact. Little or no evidence of either the nylon raincoat fibers or the wool sweater fibers are detected on the opposite item. Smooth, closely woven monofilament nylon fibers shed little and therefore have a low rate of transfer. Wool sheds and transfers readily but may not persist well on the nylon raincoat fabric. Particles of facial cosmetic powder shed and transfer readily and have a high rate of persistence, but once isolated, they may be difficult to detect.

8.5.3.1 The Problem of Detection

The problem of detection can usually be solved if the scientist knows what to look for. This is illustrated in Case Example 4.9, in which microscopic particles of paint spray were detected only by using a high-magnification reflected light microscope. In the example of the wool sweater with facial cosmetic powder on the collar, the clothing examiner can collect debris from the raincoat using a tape lift or a sticky-note lift, then search for traces of facial cosmetic powder in the collected debris.

If there is no suspect and no comparison samples are available, how should an investigative search be conducted? For example, when examining the clothing of skeletonized remains, when the victim has not yet been identified and there is no suspect, no cause of death, and no information about what transpired, it might be useful to know whether there was sexual activity involving condom use, and a finding of condom lubricant might be a useful clue. How would the clothing examiner conduct a general search that would result in the detection of a condom lubricant?

The focus of an initial investigative search should be to provide information to aid the investigator in identifying the victim, finding a suspect or narrowing a field of suspects, and locating any prior sites or transport vehicles. A search for possible condom lubricant would not have high priority. However, as the case progresses, the details of the victim's death assume greater importance and the focus of examinations shifts. If the same person is suspected of several homicides and a condom wrapper bearing a latent print is found with another victim, traces of condom lubricant in the first case may provide a link, and finding such traces assumes higher priority. Similarly, if a suspect is found to be linked to the victim, emphasis shifts from the investigative work of finding a suspect to work that is probative — either supporting, elucidating, or casting doubt on the suspect's involvement in the crime.

Detection of traces, debris, stains, deposits, and effects of damage on clothing items begins with a visual examination, at first with the unaided eye and followed by low-magnification examination using a stereomicroscope. Oblique (sideways) illumination enhances the detection of many traces. Fluorescent materials can be detected using alternate light sources, including ultraviolet and infrared light, a variable-range "forensic light source" that is used with barrier filters that screen out incident light, and light boxes designed for use in examination of questioned documents. However, it is sometimes more effective to detect materials once they have been collected and removed from the garment.

Debris can be collected manually using forceps, by physical lifting of deposits using a scalpel or spatula, by lifting with clear sticky-tape or sticky-notes, by dislodging from the substrate by tapping the fabric or scraping, and finally, by vacuuming. Vacuuming is extremely disruptive and should be used only when other methods have been exhausted. Small samples can be collected on an adhesive-tipped needle probe (Teetsov, 2002) or on a small portion of clear tape, as described in Chapter 4. Each layer of processing is more disruptive of the original deposits and should not be performed until deposits have been thoroughly documented and less disruptive collection methods have been used.

Debris, once collected, can be sorted into categories that can be described and classified. Package fibrous materials separately from particulates, plant parts from clothing and carpet fibers, and hairs from fibers. Sort fibers by color and particulates by size then type. Testing decisions should flow from the questions appropriate to the stage of the case, whether investigative or probative.

Sieving (i.e., sorting by size) can assist with detection. Each sieved fraction of dust and debris can be searched more effectively than sorting through undifferentiated debris that includes smaller particles adhering to larger ones, and undifferentiated debris biases a search toward larger and mid-sized particles (Cwiklik, 1982); see Figure 8.16. Search sieved debris using a dissecting needle or forceps to move material from a small pile

Figure 8.16 Debris collected from a pocket and placed on a sticky-note is transferred to a set of sieves. The large fiber pill was allowed to remain on the sticky-note.

of debris to a new pile a small distance away, rapidly moving particles across the space between the piles, where they are exposed to view. The target particles are either shunted to a separate pile or immediately collected. This technique is particularly effective in searching for paint chips and glass shards in hit-and-run vehicle–pedestrian collisions and is generally applicable. Pollen, tiny spheroids resulting from heat damage, and paint spray are often found in the smaller sieved fractions and may be difficult to detect in unsorted aggregate debris.

8.5.3.2 Evaluating Transfer and Persistence

Little or no evidence of transfer of component materials, such as the fibers constituting the garments suspected of contact, may mean that the component fibers do not transfer readily or, once transferred, tend not to persist. In other cases, materials are detected that are like the component material of a potential source, but the ratio of the materials is different from that found in the potential source. Examples include differential transfer of fibers in blended fabrics (Parybyk and Lokan, 1986; Salter et al., 1984) and differential retention of smaller components of transferred soil, as discussed earlier in this chapter.

Transfer can be evaluated by conducting a few simple shedding tests (Grieve and Biermann, 1997), pressing or rubbing the fabric in question against a clean piece of reference fabric with a fiber type and weave construction like that of the clothing item in evidence. The scientist can evaluate persistence by attaching the test fabric to an item of clothing during a documented period of wear. A section of the test fabric should be covered with clear tape to preserve a reference portion of the transferred fibers in their original state.

Differential transfer and persistence can also be observed on different parts of the same garment. Debris may be preferentially retained on protected parts of a garment, such as pockets and pants cuffs, or in areas of greater adhesion. Such areas may themselves be deposits, such as tar and mastics. This is illustrated in Case Example 8.12, in which debris adhering to a spot of adhesive residue persisted on a suspect's tennis shoe several weeks after a murder. Traces, debris, and bloodstains may persist on stitching threads or elastic longer than on the main part of a garment. This is illustrated for blood in Case Examples 4.4, in which bloodstains on the arm stitching of a leather coat and the side sole stitching of a pair of athletic shoes appear to have survived washing and laundering. Protected loose deposits of traces and debris, although unlikely to survive laundering, may survive extended wear.

If a fabric exhibits poor retention of debris, it may not be possible to reach a conclusion regarding contact or even lesser degrees of association. Association other than direct contact may include contact with other clothing worn by the same person, with another person in the same household or workplace, or with a site rather than a person. However, if persistence of other debris is high, transfer of non-component debris can be used to evaluate whether two items of clothing were in contact or can otherwise be associated despite the absence of component fiber transfer.

Finally, deposits of the same material may appear different on different types of fibers and fabrics (e.g., cotton twill versus polyester twill or cotton twill versus cotton corduroy), whether because one type of fabric better separates components of a stain or because a

(a)

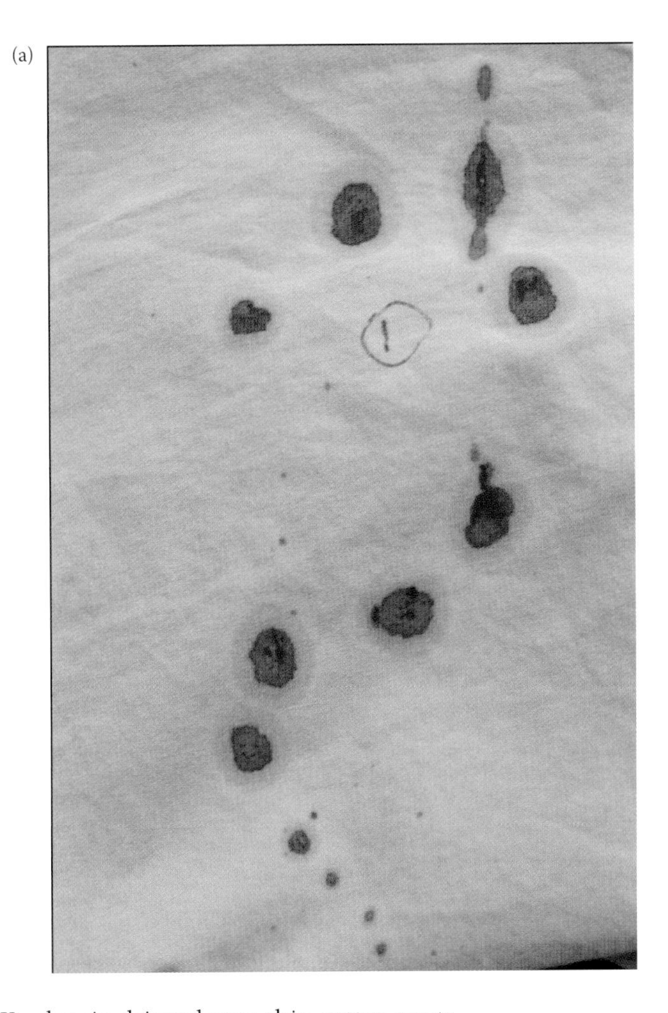

Figure 8.17 (a) Wood stain dripped onto thin cotton pants.

material deposited as a liquid or suspension exhibits different absorption properties of the liquid portion (Figure 8.17).

8.6 Questions That Can Be Addressed by Examinations of Traces and Debris

Trace evidence is considered associative evidence; i.e., evidence that can establish links between persons, objects, and locations. It is important to know exactly what is being associated or linked and to ask more specific questions:

- Does this object belong here?
- How did it get here?
- Does it belong to this individual?
- How long has it been here?
- Were these items or persons in contact?

(b)

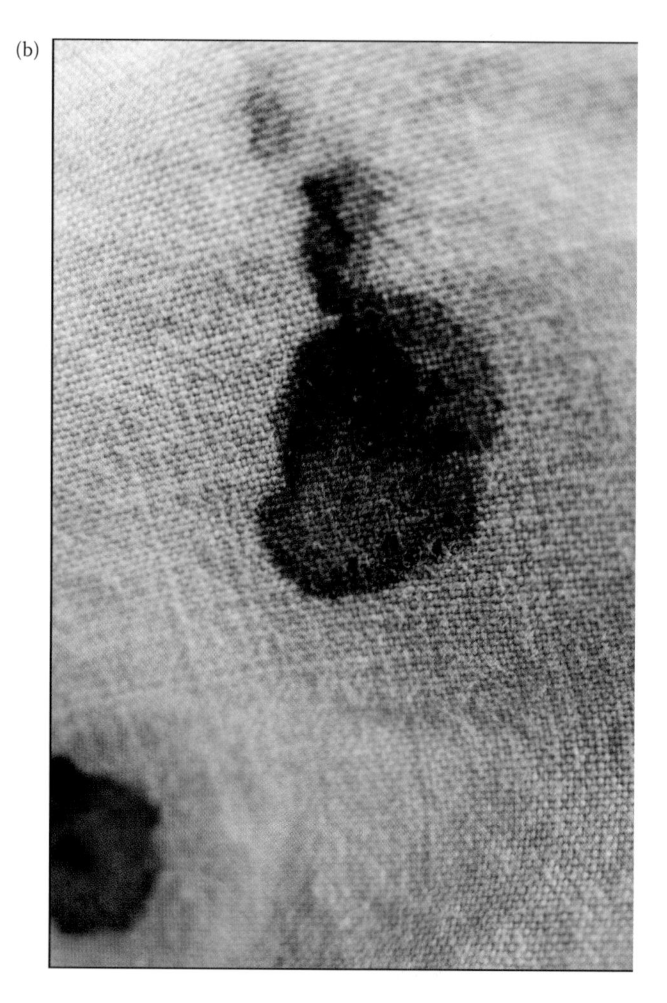

Figure 8.17 (Continued) (b) Wood stain dripped onto blue jeans.

- What is the sequence of deposits?
- What path do the traces indicate?
- In what order did things happen?
- Does it fit with the story?

The clothing examination allows the examiner to make the observations that, together with the case circumstances, form the basis for these questions.

The high specificity possible with nuclear DNA testing often allows the examiner to address the question of "who?" via analysis of body fluids and tissues. However, not every case involves DNA evidence and not every DNA test yields a highly specific result. Moreover, "who?" is only one of many case questions. Investigators also want to know what the person did, when the person left the DNA being tested, and if the person's presence at the crime scene necessarily implicates him or her in the crime. Sometimes it is important to know whether the incident was really a crime or another type of incident (Cwiklik and Gleim, 2009). A study of patterns, damage, and trace evidence can often address questions that cannot be addressed by DNA evidence alone.

CASE EXAMPLE 8.10

Does this object belong here? Does it belong to this individual?

Police said that a penknife found on the pavement near the hand of a man who was shot had been used to attack or threaten the policeman who shot him. Debris in the blade well of the knife was compared with debris in the pockets of the deceased, but they did not correspond — the debris in his pockets included particles that indicated he was a nibbler and a smoker, but no food or tobacco particles were found in the knife. Instead, the examiner found a different set of debris in the knife, including wood shavings, several types of paint particles, and numerous types of fibers. These particles and fibers corresponded with debris in the police officer's pockets — he was a do-it-yourselfer and did home carpentry — establishing a link with the officer. Because the correspondence was nearly complete, the knife was linked not only with the officer, but also with his uniform pocket, establishing that the knife had been carried by the officer in his uniform pocket and "planted" at the scene near the hand of the deceased. (Cwiklik, 1999, Case Example 1)

CASE EXAMPLE 8.11

Were these items or persons in contact? In what order did things happen?

Case Example 4.7 describes a case in which a woman was killed and her body found outside a 400-unit apartment building. The debris on the victim's jersey knit dress exhibited a high degree of correspondence with the debris on one side of a blanket found in the apartment dumpster and a lesser degree of correspondence with the other side of the blanket. The high degree of correspondence of numerous types of particles and debris, and only a few types of lint particles that did not correspond, established that the victim had been wrapped in the blanket. Debris on the dress and blanket included gold-colored carpet fibers, numerous green indoor–outdoor carpet fibers, and many other types of fibers and particles.

Before the body was discovered, a suspect had drawn the attention of police when he tried to set a fire in the dumpster. By the time the body was found, he had moved out of the building and the apartment manager had cleaned the man's vacant apartment and rented it to another tenant. However, just before the suspect moved out, he had a garage sale, selling his furniture and household goods to fellow apartment dwellers. In a search for potential sources for the predominant types of fibers on the dress and blanket, thus a potential link with the perpetrator or the murder scene, police went to each unit in the apartment building looking for the dispersed household furnishings of the suspect. They obtained reference fiber samples and brought them to the laboratory. Fibers from the carpets and upholstered furniture sold to the neighbors corresponded with predominant debris fibers on the victim's clothing and the blanket in which she was wrapped. The gold-colored fibers were linked to a small rug sold at the garage sale. However, the green indoor–outdoor carpet fibers remained a mystery.

Police continued to inquire among neighbors and friends and eventually found indoor–outdoor carpet that had belonged to the suspect. He had it only a short while, having bought it after his wife left, not long before the incident under investigation, and then selling it when he moved out. The presence of these fibers on the victim's clothing established a time frame for the murder. The presence of the set of fibers that reflected the suspect's household furnishings provided a strong albeit not definitive link with the suspect and the site. This allowed police to obtain a search warrant to the apartment and find the bloodstains discussed in Case Example 4.7. The murder was linked to the suspect's former apartment by virtue of bloodstain patterns on the apartment floor that corresponded with bloodstains on the blanket from the dumpster (Cwiklik, 1999).

CASE EXAMPLE 8.12

What is the sequence of deposits? What path do the traces indicate? How long has it been here?

The body of a woman was found in her apartment near an overturned potted plant, with soil outlining the position of the body when it was removed. Blood was observed under the body, but not much blood was shed overall. Baking flour found in a sack on the kitchen counter was spilled on the kitchen floor and tracked to the front door. There was neither an obvious suspect nor an obvious motive. Was the perpetrator discovered while ransacking the apartment? Was the spilled flour and overturned plant evidence of a struggle? Samples of the flour, the soil from the potted plant, and control samples of carpeting, upholstery, and several blankets were collected. The victim's clothing and swabs were submitted to the laboratory. Investigators found no evidence of sexual activity on any of the swabs or on the victim's clothing. No foreign hairs were observed other than types found in the background debris of the apartment. Soil was found on the victim's clothing, mostly on the front, but no white flour was observed. This suggested that the potted plant was overturned after the victim was no longer able to move and the sack of flour was spilled in the kitchen afterwards, possibly in a search for valuables.

A suspect was apprehended several weeks later. Police submitted his shoes to the laboratory. No blood was observed. In the intervening weeks, debris had been lost and added. However, there was a small rectangular sticky spot on one of the shoes where it appeared that a piece of sticky-tape had been on the shoe and fallen off. Adhering to the sticky spot were several fibers. The fibers were correlated with fabrics and carpet in the victim's apartment: gold-colored carpet, a brown blanket, a blue blanket, and a multicolored knit afghan. A dusting of flour, consisting of wheat starch and a smaller number of milled wheat particles, was also observed on the sticky spot. The dusting of flour reduced the available surface of the sticky spot, preventing the adhesion of additional debris. The examiner found a few other types of fibers and particles under the carpet fibers, but not many. Thus, the deposit represented a narrow window of time and reflected not only the predominant fiber sources in the apartment, but also the flour that was spilled during the incident. This linked the suspect not only to the scene, but also to specific activities that were a part of the crime.

CASE EXAMPLE 8.13

Does it fit with the story? Absence of deposits significant

A 15-year-old young woman told the police that she had been dragged into a field across from her apartment building by a man she was able to name. A man with that name lived in her building. She said that they tussled in the grass, where he raped her. An examination of her clothing showed seminal fluid stains in the crotch of her blue jeans, but all her clothes were otherwise remarkably clean. The absence of the plant material, soil, and grass stains that would be expected from a tussle in the field suggested that events did not transpire exactly as reported. The laboratory promptly conveyed results to detectives. When confronted with this information, the young woman admitted that she was afraid she was pregnant after unprotected intercourse with her boyfriend, told her mother she had been raped, and filed a report when her mom insisted. She made up a common description for her supposed assailant (Caucasian, medium build, light brown hair) and a common name. The man with that name who lived in her building also fitted the description. He had been detained by police for questioning, but because of the laboratory findings, no charges were filed.

8.7 Questions of Contact

The non-component debris found on an item reflects the individual history of that item. It thus has the potential for uniquely identifying the item and, when transferred, for providing information about contact that is far more specific than would otherwise be possible. This potential depends on whether there are enough particles and fibers, there are enough types of particles and fibers, and the number of corresponding particles and fibers in the sets of debris significantly outweigh the types of particles and fibers that do not correspond (Cwiklik, 1999). The potential for establishing contact via comparison of sets of debris is most likely to be realized if the clothing was put into evidence soon after the suspected contact or if it was not laundered or heavily used between contact and collection. The reference point is subsequent use rather than elapsed time.

The scientist can decide during the initial clothing examination whether conditions for a context-based debris comparison are likely to be met. This can be done by collecting a few preliminary samples on clear sticky-tape and examining them under the stereomicroscope, as discussed in Chapter 4. It may not be possible to definitively establish contact via comparisons of debris if the debris populations of samples from two sources (for example, a suspect and a victim) fail to exhibit high correspondence at this preliminary stage. If the preliminary samples show promise of high correspondence when examined under a stereomicroscope, they should next be examined at higher magnification to confirm a likely microscopic correspondence. This preliminary examination is not sufficient to permit conclusions, but is a useful tool in deciding on an analytical approach.

If insufficient information exists to establish contact, the examiner may perform a target search. A target examination begins with a search for fibers like the component material of the object that may have been in contact with the item being examined. If the question is whether two individuals were in contact, a target search of clothing would focus on component fibers. Even in a target search, if the scientist finds predominant debris fibers that the garments from both persons have in common, these should be reported and source items found if possible (Cwiklik, 1999; Deadman, 1984a, 1984b; Lowrie and Jackson, 1994; Taupin, 1996). This is illustrated in Case Example 8.11. Debris on the victim's clothing included numerous fibers of a number of types that corresponded with carpeting and upholstery fibers traced back to the suspect's residence. This correspondence provided a strong association, but it was not known from this finding alone whether the association was with the site, the suspect, or the blanket. Someone else could have taken the blanket from the apartment or the suspect's car, then killed the victim elsewhere. However, the fibers together with blood pattern evidence provided a firm link with the site.

Items from a suspect are most often of interest only if they are linked with the incident under investigation. Initial examination of a suspect's clothing is usually a target examination. If the component fibers of the victim's clothing, fibers like the major fabric sources at a site, or particles that characterize the particular event are not found, there is seldom a reason to examine the remaining debris. The same considerations apply to a suspected site. If the suspect's clothing does not retain much debris, it may be that some of the predominant types of debris from a victim are retained better than the component fibers of the victim's clothing. In this case, the examiner can conduct a modified target search for preliminary evaluation.

8.8 Target vs. Context-Based Examinations

Target examinations are based on a search for a specific type of foreign fiber, particle, or stain that would be significant if found. These would usually be the component material of the item of interest. For example, if the clothing of an assault victim includes a bright purple cashmere sweater, it would be reasonable to expect fibers from the sweater to transfer to the perpetrator. One might then search among items from a suspect for corresponding purple fibers. A garment bearing such fibers would warrant further examination. If bloodshed had occurred, one might search for blood. Modified target examinations include searches for major types of foreign fibers and particles as well as for component fibers.

A context-based examination involves a study not only of transferred (foreign) component fibers, but also of the non-component foreign debris that would accompany them (Cwiklik, 1999).

Sampling techniques for target and context-based examinations will be discussed later in this chapter.

8.9 Absence of Debris

Suppose you are trying to determine whether a beer bottle used in a crime had at one time been on a particular shelf. If the shelf is dusty and there is a circular-shaped void in the dust, the void is evidence that something, possibly the beer bottle, had been there and was removed. Other objects with circular bases of that size and pattern are alternative possibilities. If the dust is undisturbed, or if only shapes unlike the bottle base are observed in the dust, it is safe to conclude that no such bottle has been there since the dust was deposited. A conclusion about the absence of the bottle depends on the presence of dust. If the shelf is clean, there is no information about the presence or absence of the bottle. Similar reasoning can be applied to an absence of debris.

If few particles and fibers are observed on a garment, either the garment is new or it does not readily retain debris. In the latter case, any transferred debris is unlikely to persist. There is no information to distinguish between lack of transfer and transferred material that has been lost.

If many particles and types of particles are observed on an item of clothing, but not the component material of the object suspected of being in contact with it, either the objects were not in contact or the component materials do not shed or transfer readily. As discussed earlier, shedding potential can be tested, and potential contact of low-shedding fabrics can be evaluated by a search for transfer of non-component debris. If neither the component fibers of a garment nor its non-component lint and debris are found on a second garment — and the second garment itself bears a normal range of debris — recent contact between the garments can be excluded. The examiner must have a basis for concluding that the transferred material would be detected if it were there.

In summary, contact can be excluded in the absence of corresponding particles under the following conditions: (a) when the probability of transfer would be high had contact occurred; (b) when the probability of persistence would be high had material been transferred; and (c) when the probability of detection would be high had material been transferred and retained. Contact cannot be excluded despite the absence of corresponding particles when the probability of transfer, persistence, or detection would be low even had contact occurred. If it is not known whether those probabilities would be high or low, the scientist might perform a few experiments to find out, using types of debris and substrata similar to the evidence materials; otherwise, no conclusion should be reached (Cwiklik, 1999).

Reaching conclusions from an absence of debris rests partly on expectations of what would be observed had contact occurred. This can be generalized to other types of associations (links) and used in reconstruction of events. For example, if an assailant was tussling with a victim in a grassy field during a sexual assault, we expect the clothing of both individuals to exhibit grass stains, soil, and other plant debris, including deposits of such materials on the inside of the underpants. The absence of these types of deposits was significant in Case Example 8.13 and resulted in police dropping charges against the suspect. In contrast, an absence of debris in Case Example 8.14 (below) did not permit a conclusion.

CASE EXAMPLE 8.14

Absence of debris; no conclusions possible

A teenaged girl did not come home after school and was brought to her home later that evening by a young man. She reported that he had abducted her, forcing her into the trunk of his car, took her to his house, raped her, and then took her home when his girlfriend was due to arrive. She did not report this to police until the next day. Police took her to the hospital, where sexual assault samples were taken and her clothing was collected. They obtained a search warrant for the young man's car. The car had been detailed (thoroughly and professionally cleaned) that morning, so little evidence of value would have remained other than reference samples of the carpet and upholstery. Her clothing was examined for any evidence of contact with the trunk carpet. No carpet fibers were observed. She had been wearing a raincoat over polished cotton shorts and a nylon top. None of these fabrics retained much debris, and carpet fibers usually exhibit low persistence. Almost all the non-component debris from the trunk had been removed by cleaning. The absence of carpet fibers on the young woman's clothing provided no information about whether she had been in the trunk.

8.10 Summary: Nature, Composition, Source, and Transfers of Traces and Debris

A residue or trace of something can provide information about the person, object, or event itself. Debris is the particulate record of processes produced over time and thus provides a historical record. The debris in an environment is characteristic of that environment. The debris on a person and on the person's clothing reflects that person's habits and environments. When two objects or beings come into contact, a transfer of material occurs (Locard's theorem). When a transfer of material occurs upon contact, the material transferred includes lint and debris.

The debris population of an item of clothing consists primarily of component fibers separated from the item itself and from other clothing worn at the same time, and a "native" population of non-component particulate and fibrous debris characteristic of the wearer's person, habits, and environments. The sets of component and non-component debris characteristic of a person or environment are referred to as normal debris. The debris population also includes a smaller set of foreign debris; that is, particles and fibrous material that result from other processes and sites and are not characteristic of the wearer. It is the latter that can be used in forensic investigations as evidence of contact or another association.

Only clearly foreign material can provide evidence of contact with another object or person. The set of foreign fibers and particles may include a high proportion or a small proportion of debris from a single contact. An expectation of one or the other is based on the case circumstances and the item history. This expectation and examination of preliminary samples will govern the scientist's decision about whether to examine target fibers only (if a small proportion would be expected) or to perform a context-based examination involving transfers of debris (when a high proportion would be expected and the expectation is corroborated by preliminary samples).

When traces or debris that might be expected are not found, or are found in different ratios on the source item, the absence of debris can be evaluated with respect to expected transfer, persistence, and detection. If specific debris would be expected had contact occurred, the absence of debris can be used to exclude contact between two items. However, if the expectation of transfer or persistence is low, the absence of debris provides no information and no conclusion can be reached.

8.11 Sampling and Sorting

Samples can be collected for context examinations, target examinations, or preliminary examinations that allow the clothing examiner to decide which type of examination to perform. Sampling rationale, criteria, and techniques are discussed below.

It is good practice to take samples of debris even if traces and debris do not at first seem to address the case questions. As new case information is developed, you can reevaluate the initial decisions about which examinations to perform. For example, if a crime was perpetrated at the residence of a victim, evidence associating a suspect's clothing with items from the victim may link the suspect with the scene but not specifically with the crime. The suspect may have been at the scene, but before the incident under investigation, or may have arrived afterward. Another suspect may be developed later. Clothing that was hidden after being used in the crime may be discovered. In one such case, bloodstained clothing was found beneath a floorboard in the residence of a suspect nearly two years after a homicide. If you collect samples of debris during the initial examination, they will be available to examine as new information suggests that they would be useful.

It is best not to assume that additional samples can be collected at a later time. After clothing is examined in the laboratory, it is usually returned to a police evidence room, where detectives, prosecuting attorneys, defense attorneys, and defense investigators may view it. Although there are usually ground rules, such as placing the items on butcher paper and wearing gloves, they are not always followed to the letter and are seldom followed with the consciousness of transfer one would expect of a forensic scientist. If the reevaluation is for a new trial after appeal of a conviction, the evidence will often have been handled in court and stored in the court evidence locker during the original trial and afterwards, often with little protection from contamination. Evidence may even be inadvertently destroyed, as in Case Example 8.9. It is best to take samples during the initial examination.

8.11.1 Sampling Rationale

The bases for sampling decisions are discussed in Chapter 4 and are repeated here in summary. Sampling decisions should flow directly from the questions in a case. The selection of samples should be governed first by significance, then by specificity and utility. Significance is defined as the potential to narrow down hypotheses. Specificity is defined as the potential to narrow down sources. Utility or usefulness is defined in this context as the potential to produce interpretable results.

If several deposits of the same type of material exist, a deposit that can address what happened is more significant than a similar deposit that does not, even if the latter may yield more specific results. This is illustrated in Case Example 4.6 and in Case Example 8.15, which follows. For traces and debris, the potential significance of isolated particles and fibers is also a factor. You should collect every potential human hair or glass shard, for example, but a sampling of adhering fibers may suffice. You may collect as many target fibers as possible (component fibers are targeted even in context-based examinations), but only a sampling of non-component debris. When sets of debris are compared in addressing questions of contact, a few isolated fibers or particles is not significant to debris comparison because more than a few are required to be part of the characteristic debris of a person or a site. However, if isolated particles or fibers are marker types for the incident under investigation, even a small number may be of interest.

CASE EXAMPLE 8.15

Significant vs. specific; reconciling witness statements

A vehicle was passing two motorcyclists driving side-by-side on a limited-access highway at night. The vehicle sideswiped one of the cyclists, sending the motorcycle and its driver across the pavement and into the path of another car. The driver of the second car stopped, but the first driver did not. Several people at a nearby bus stop saw the first vehicle and described it as a black car. Police submitted the victim's clothing to the laboratory for clues to the make and model of the first vehicle.

The motorcycle fender and the left leg of the victim's pants exhibited turquoise blue paint smears. Paint chips were also collected from the motorcycle and the victim. Because the incident occurred on a busy thoroughfare, the paint chips were of several types, with five or six different topcoat colors represented. Among them were turquoise blue paint chips (the topcoat color was like the color of the paint smears), white paint chips (which were correlated with paint from the second vehicle to strike the victim), and paint chips of several other topcoat colors, including several black paint chips. The turquoise blue paint smear was the most significant, having been deposited by impact. However, the paint chips were more specific, exhibiting undercoats and layer structure that could better narrow down a potential source vehicle. With the turquoise blue smear as a reference, paint chips with a turquoise blue topcoat were selected for further testing. The other paint chips were reported to detectives but not further examined.

Should police continue to search for a black car? Sodium vapor lamps lighted that particular stretch of highway. How could this affect witness descriptions of the vehicle color? Investigators shone a variable-wavelength light source onto the blue paint smear on the pants. The turquoise blue paint appeared black at a major wavelength of sodium emission, an orange color complementary to turquoise blue, explaining why the witnesses described a black car. Upon hearing this result, detectives said that the other cyclist, who would have seen the vehicle by the light of his headlamp, described a blue car.

8.11.2 Sampling Criteria

General sampling criteria are discussed in Section 4.4.1 and are summarized here. An adequate sample is of sufficient size to permit the requisite testing and of sufficient quality to permit clear interpretation. A sample of sufficient quality truly represents the material of interest and does not include anything extraneous.

For traces and debris, searches for specific types of fibrous or particulate materials must be broad enough to include atypical specimens. For example, a search for hairs will yield all hairs on an item only if fibrous materials that may or may not be hairs are collected as well. Some hairs exhibit atypical morphology even under a stereomicroscope, especially if both ends of a hair exhibit broomstick fractures. These hairs may look more like plant parts than hairs during a preliminary search. Therefore, initial sampling for hairs should include fibrous materials with that appearance, even though most of them will probably be plant parts or fibers (Figure 8.18). Similarly, sample collection targeting human hairs often includes animal hairs. Sample collection for glass fragments often includes quartz granules, which, like glass, exhibit conchoidal fracture. Subsequent sorting using crossed polars permits the highly birefringent quartz particles to be distinguished from glass, which is not birefringent or exhibits strain birefringence only.

A sampling for hairs with roots may include hairs with abraded broomstick fracture having the shape of a telogen (shedding phase) root. Including a broad range of characteristics in initial sampling ensures that sufficient numbers of the fibers and particles of interest are collected. In other words, sensitivity is more important than specificity in the initial search. Although a good sample does not include extraneous material, each

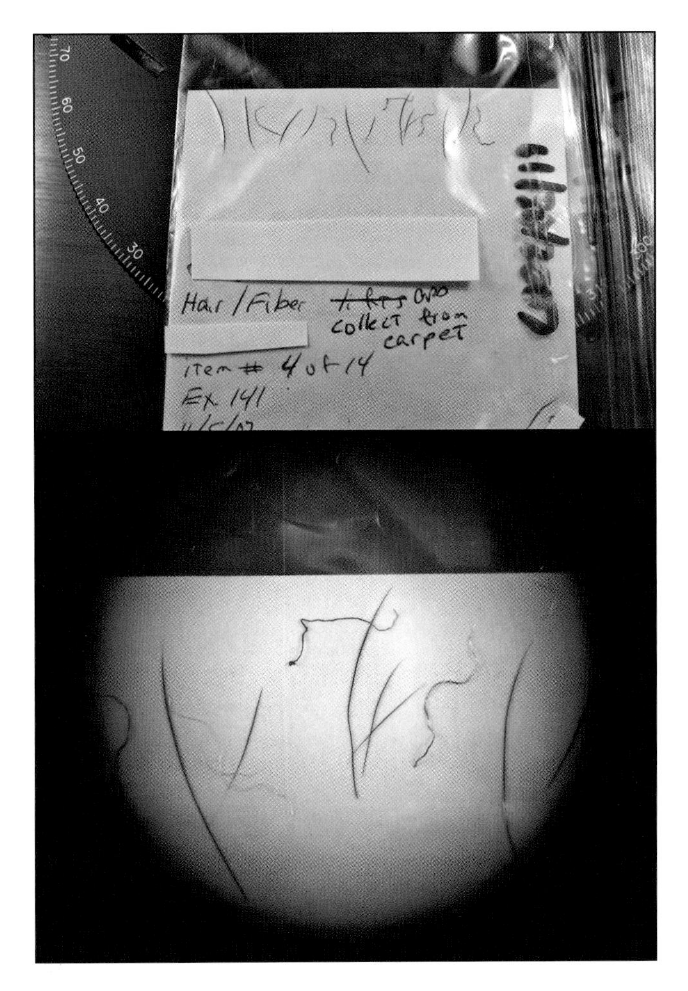

Figure 8.18 (Top) Possible hairs collected onto a sticky-note. (Bottom) Initial sampling for hair includes fibers (6.4× magnification).

particle and fiber can be individually distinguished using microscopy and is not compromised by another separate particle or fiber. The sample can be further sorted at a later stage.

Other sampling decisions for traces and debris involve preliminary sampling to determine whether a target-based or context-based examination should be conducted and whether to sample everything or only the most recent deposits. If the clothing was collected or removed from use not long after the incident under investigation, a reference sample of accumulated debris should be collected, but only recently deposited foreign debris. In most cases, recently deposited debris consists of surface debris, not debris that has worked its way into the fabric. However, particles such as glass shards and smaller paint chips may penetrate deeper into the fabric even if recently deposited. Target searches for glass and paint may require not only manual collection using forceps and needle probes, but also gentle tapping or shaking of the item to release loose particles, even scraping the item to dislodge the debris. The latter methods should be used only after other types of evidence such as bloodstains and gunshot residue have been sampled and damage analysis has been completed. Shaking is done with the clothing item held inside a large paper sack kept closed during the process. The bag is opened only after the dust and debris have settled to the bottom. The dust and debris can then be sieved to facilitate a search for glass or paint.

Must traces and debris always be collected? Traces and debris do not address every case question, nor are they significant in every case. However, one can rarely be certain of this at the outset, or even well into the examination. Suppose that a body of a woman who was shot is found in her home, and the suspect is her husband. A finding of hair and debris transfer would mean little to corroborate his involvement, because this would also be expected in the normal course of events, even if someone else was the perpetrator. Not only could someone else be the perpetrator, but there might be event-specific debris even if the husband was the perpetrator. Suppose that a stray bullet was fired into the ceiling, producing plaster and paint dust that fell onto the clothing of the victim and the perpetrator. This would be marker debris characterizing the incident, thus highly significant.

How can the clothing examiner perform thorough examinations yet still make reasonable decisions about effective use of time? With respect to sampling, evidence such as traces and debris that can readily be lost, contaminated, or otherwise compromised should be collected even if follow-up examinations are not anticipated at that time. In the latter case, the examiner should use a rapid sampling method that permits rapid follow-up screening. Collecting traces and debris on sticky-notes or clear sticky-tape, although neither optimal nor convenient for many types of follow-up testing, accomplishes the preservation of evidence. Placing the sticky-notes in clear plastic bags or the clear tape on plastic document protectors allows for rapid screening. The collected debris should be screened before it is packaged and sealed in case something unexpected requires attention.

8.11.3 Sequence of Sampling and Collection

The scientist who first examines an item of clothing is responsible for determining a logical sequence of sampling and testing that may involve other forensic disciplines. It may be necessary to confer at the outset with scientists in those disciplines. Consider clothing submitted in a homicide masked as arson. Body fluids, trace evidence, firearms-related deposits, damage, and flammable liquids may all be of interest, but optimum handling for one type of evidence may compromise another type. If the clothing is stuffed into an airtight can for analysis of flammable liquids, then bloodstains, gunpowder particles, and damage characteristics may be compromised. If the bloodstains on the clothing are allowed to air dry, the flammable liquids may evaporate. It is urgent for the examiners to confer immediately so that selected bloodstains and other deposits can be excised prior to the clothing being preserved for flammable liquids testing. In other examples, a ligature may be tested for traces of DNA but also for trace evidence. The smooth jacket buttons of an assault victim whose clothing was removed during an assault may require dusting for latent prints as well as swabbing for DNA and collection of trace evidence. Chemical mapping for gunshot residue, saliva, and seminal fluid removes trace evidence and alters characteristics of deposits and damage, which should thus be examined first. Even if the clothing is submitted to one section of the laboratory for specific types of examinations, you must safeguard other types of evidence for their potential information value should new questions arise as the case progresses.

As mentioned earlier regarding shaking and scraping clothing to dislodge debris, sampling methods that may disrupt other deposits or patterns should usually be performed last. The exception is clothing encrusted with mud or plant debris that obscures other stains, deposits, and traces. Encrusted material must be removed to permit examination of other evidence. Large crusts can be removed with gloved hands and placed on clean paper. The side of the crust that was closest to the garment can then be examined for traces and debris. If the encrusted mud or sand is loose and crumbling, the examiner can allow the loose soil or sand to fall into a paper sack and then gently dislodge what remains manually with a spatula. Think of emulating a field archaeologist, who carefully and gently removes material layer by layer hoping to expose embedded artifacts.

8.11.4 Techniques for Sampling and Collection

"Trace evidence recovery or collection techniques used should be the most direct and least intrusive technique or techniques practical. Collection techniques include picking, lifting, scraping, vacuum sweeping, combing, and clipping" (SWGMAT, 1999). Guideline 5.4 of the SWGMAT trace evidence recovery guidelines appears in Section 8.15. Additional topics are discussed below.

8.11.4.1 Sample Size and Composition

The examiner should leave enough material on the garment to permit reexamination of a stain or deposit in context, but collect enough to perform the desired examinations. Inhomogeneities and outliers must be included in a sample if included in the deposit.

8.11.4.2 Sampling and Sorting Techniques

Sorting: When aggregate debris includes fibers and particles, a range of sizes, various colors, and materials of different types, such as plant parts and fragments of synthetic materials, then preliminary sorting becomes part of the sampling process.

Segregation of gross types of material: Unless the sample size is very small, target searches and preliminary screening are faster and more thorough if preliminary sorting is performed during the collection process. Hairs, fibers, plant parts, and miscellaneous fragments such as paper and plastic can each be placed in separate clear plastic bags. Particulates can be placed onto a piece of paper, which is then folded in a double three-way fold (a bindle) or be collected on sticky-notes.

Sieving is performed using brass sieves. For traces and debris, 300, 140, and 60 mesh sieves and 0.5 and 1.0 mm perforated disk sieves are useful size ranges. Sieves of larger size range are useful for screening out twigs and roots, coins, buttons, pebbles, broken fingernails, and so on. For larger materials, a kitchen strainer will do if large sieves are not available. Sieving should be used in target searches for glass shards and paint fragments (Cwiklik, 1982) and for finding bits of burnt clothing and related artifacts in ashes from a burn barrel or fireplace (a suspect may have burned clothing worn during commission of a crime); see Figure 8.19. Sieving may be used for evaluating whether a dusting of soil on clothing might correspond with the smaller size ranges of a potential source. Sieving can also be used to draw a histogram (distribution graph) of particle sizes, which can sometimes be distinguished by proportions of components of a material or particle sizes of components. The latter is exemplified by fuses sold as lighting for roadside emergencies (Suzuki,* personal communication), found in one case as traces on clothing. Histograms are also used to characterize soil.

* Suzuki, E.M., Washington State Patrol Crime Laboratory in Seattle.

Figure 8.19 (Top) Zipper teeth loosening during a test burn as supporting fabric disintegrates. (Bottom) Zipper pull and zipper teeth falling beneath grill in a test burn. Sieving of burn barrel debris allows recovery of such artifacts.

Color sorting of lint fibers: A 2″×3″ microscope slide can be prepared with mounting medium (distilled water or a suitable refractive index liquid) in six areas, each for a different color group (Figure 8.20). As you remove fibers from the clothing, or from a sample already collected from the clothing, you can mount them directly in one of the six areas of mounting medium. This permits red fibers from one item, for example, to be rapidly compared with an assortment of red fibers from another item (Murren,* personal communication).

Particle-picking with forceps: Target particles of fibers can be collected with forceps or a needle probe. For example, the active controlled substance in a capsule may be sampled for further testing by particle picking (Murren, personal communication). This method is applicable to traces and debris found on clothing. The scientist can use forceps to isolate paint chips found in a smear of grease, metal shavings found among sawdust, or fragments of fiberglass resin found in plaster dust deposits.

* Murren, C.A., Forensic Services, Seattle, Washington, formerly with the Washington State Patrol Crime Laboratory in Seattle.

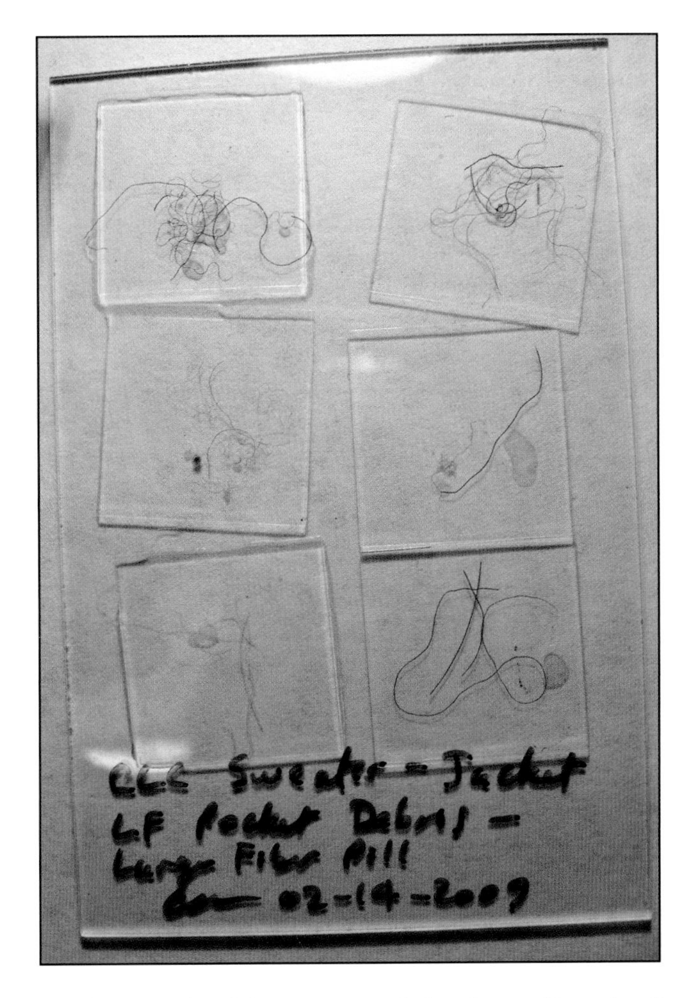

Figure 8.20 Fiber pill from pocket debris was sampled and sorted by color directly onto microscope slides.

Particle-picking with needle probe: Particles and fibers adhere to needle probes by static electricity; thus, it is a good way to collect them from clothing or samples of debris. Larger particles can be manipulated with a dissecting needle. Smaller particles can be collected using a sharpened tungsten needle. The needle is carefully touched to the particle, causing the particle to adhere to the needle and allowing it to be transferred to a microscope slide. The transfer should be observed with a stereomicroscope. If the particle is to be mounted, it can be placed directly into a drop of mounting medium, or a micro-drop for very small samples. Needle probes are well suited to collecting such small amounts of material that the evidence is barely altered (Teetsov, 2002). It is thus ideal for preliminary sampling to determine whether additional testing is warranted and what type of testing to perform.

Needle probe with mounting medium: Some types of particles tend to fly off when handled with forceps, especially glass and other hard materials such as mineral grains, chunks of hard plastic, and brittle paint chips. An alternative technique for collecting individual particles of glass and other hard fragments of material is to touch a needle probe to a mounting medium, allow the glass shard or mineral grain to adhere to the liquid, and then transfer it to an adjacent microscope slide already prepared with the same mounting medium. Refractive index oil is used for particles that are to be mounted in the same medium. Distilled water is a temporary mount that evaporates, used for water-insoluble particles that are to be examined dry or that require preliminary examination before selection of a more permanent mounting medium. Particles can also be mounted in a medium that can be rinsed off with solvent using micro-techniques (Teetsov, 2002).

Sticky-tape: Touch a piece of clear sticky-tape to the deposit using a hard, narrow object such as the tip of a retracted ballpoint pen. This is a quick technique that allows a small deposit to be rapidly collected for preliminary examination using a stereomicroscope when using a needle probe is not convenient. Preliminary examinations with a compound polarized light microscope can also be conducted with the taped sample. Place the tape on a microscope slide, add mounting medium, and cover with a cover slip. The clear tape should be selected for low retardation to minimize interference with sample birefringence.

Sampling powders with a needle probe: A bit of adhesive, such as collodion or adhesive from water-soluble tape (Teetsov, 2002), can be placed on the tip of a needle probe. The adhesive-laden needle tip can then be swept across powdery deposits, thereby collecting samples that can be transferred to a microscope slide.

Sticky-note lifts: A light taping of the garment using sticky-notes or gel lifts can supplement manual collection of visible particles and fibers. For hairs and larger fibers, larger sizes of sticky-note are now available in office supply stores. After collecting traces and debris, place the sticky-notes in clear plastic bags such as snack bags or sandwich bags to permit rapid screening and preliminary examinations. Sticky-notes are also convenient when it is less important to collect a particular particle and when it is sufficient to collect a sample of a multiparticle deposit; although for particles that are very small or fine, subsequent removal from the sticky-note may be difficult, and this method should not be used with powders.

Tape lifts: Clear tape can also be used to supplement manual collection of particles and fibers. Several methods are in common use: clear wide tape to lift debris; clear tape on rollers; large pieces of clear tape with printed grids; and gel lifts. The examiner must decide whether the location of the particles is important. If not, the tape can be touched to the garment several times. If location is important, adjacent pieces of tape should be placed on the garment and gloved hands rubbed over them. In either case, the tape with debris is removed and placed on a document protector or other clear plastic that permits *in situ* examination using transmitted light. Individual particles, hairs, or fibers are removed with forceps or a needle probe, usually with a drop of solvent, after a small slice has been made in the tape or after it has been carefully lifted from the plastic backing.

Sometimes, in the initial evaluation of how laboratory work can address the case issues, it is obvious that trace evidence examinations are likely to be important. In that case, stains, deposits, traces, and debris should be examined with a stereomicroscope and trace evidence collected manually. However, if at first it seems that the case questions would not be addressed by trace evidence, sticky-note lifts or tape lifts are a valuable method to collect traces and debris just in case they will be needed as the case progresses and new information is developed. Collecting traces and debris on tape lifts or sticky-note lifts alters the evidence in that the material removed can no longer be evaluated for manner of deposit or for co-deposit with other particles and fibers. Furthermore, the adhesive must be removed from the adhering particles and fibers for many types of follow-up examinations, potentially damaging some particles and fibers. However, in most cases it is better to have a non-ideal sample than to have one that is contaminated or lost because of a failure to collect traces and debris during the initial examination.

8.11.4.3 *Special Problems in Sample Collection*

Heavily encrusted material: The encrusted material, usually soil, must be removed to permit examination of the garment, using the techniques discussed in Section 8.11.3. The encrusted material that has been removed must be preserved. If the person wearing the clothing was buried, or if the clothing itself was buried or concealed, small objects or artifacts may be embedded in the soil and can be recovered by carefully and gently crushing the collected encrusted matter and sieving the resulting loose material.

Loose powders: Loose powders result from crushing encrusted deposits as described above. In addition, sand, sandy soil, and other materials that do not readily form clumps may be encrusted on clothing items when first recovered, but become loose deposits as they dry. Sand and other deposits composed of relatively large particles can be allowed to fall into a paper bag (gently shake the garment to facilitate removal) and then should be transferred to another container for screening and examination.

Clay soil usually forms crusts, but if the crusts break up, fine loose powder may result. When the garment is removed from the container in which it is packaged, a small dust cloud may rise. Not only does the rising dust contaminate the work area, but it may also be a health hazard if inhaled. The examiner should immediately close the container and put on a mask before proceeding. If a HEPA filter is available, it should be placed nearby. Several lengths of paper towel can be moistened and placed across the open sack and allowed to remain until the dust has settled. The garment can then be carefully transferred to a very large clean sack (the size that is usually available in hardware and gardening supply stores), gripped at the top with the bag closed, and gently shaken. The garment can be secured in place with a clip and the bag hung until the dust has settled. This may take several hours. The scientist may then remove and examine the garment as usual. The dust can be transferred to a smaller container, or several small containers, and examined for traces and debris. In regions of low humidity, it may be helpful to mist the air with water from a spray bottle before removing the garment or the dust.

Ashes: Clothing is sometimes burned to destroy evidence, and recovery of unburnt or partially burnt remains may involve sifting of ashes from a burn pit, burn barrel, or fireplace. Like fine powders, ashes easily rise and remain suspended in the air. Facemasks, HEPA filters, moistened paper towels, and spray bottles for misting dry air are similarly useful. However, unlike fine soil or dust, ashes tend not to settle. A good way to sort through ashes is to sieve them using a sieve set with a tight lid. Objects and fabric fragments usually remain in the larger mesh sieves. The finer sieved fractions that are mostly ash, if plentiful, should be divided into small portions before examining for traces and debris.

Feathers: When a feather pillow, down jacket or vest, or down sleeping bag has been damaged by bullets or a knife, small feathers tend to fly out during examination. The damage margins must be examined but the feathers must be contained. The problem is to cover the defects without disturbing the defect margins. One technique is to affix a small piece of clear plastic over each defect without touching the defect margins. Use clear tape. Do not place tape directly over the defect. Preliminary examinations can be conducted through the pieces of clear plastic, and for detailed examination, the plastic can be removed one defect at a time and replaced after the examination is completed.

Embedded dust: Fine dust and debris that is carried by the wind in some regions and some occupational debris can become embedded in fabric and may survive a casual brushing. Most such material is usually recovered by gently tapping the underside of the fabric. When insufficient sample is collected by tapping, or there is some danger associated with breathing the dust, vacuuming the clothing item may be indicated. (See the SWGMAT guidelines in Section 8.15.)

8.11.5 Collecting Samples for Target Examinations

A target examination should be thorough. Obviously, it is important to search thoroughly before deciding that no evidence exists that may link a suspect with a crime. There might be traces that persist under the least likely circumstances: on a type of fabric that retains little debris, on a garment that has continued to be in use after the incident under investigation, and sometimes even on garments that have been laundered or dry-cleaned. Case examples of surviving bloodstains found on the stitching threads of a jacket that had been hand-washed and on the threads on the side soles of an athletic shoe washed in a washing machine are presented in Chapter 4, and a case example of debris surviving on a sticky spot on a suspect's shoe is presented in Case Example 8.12. Particles and fibrous materials may be loosely caught or entangled or protected from subsequent movements across the fabric surface that might otherwise remove it. Protected areas include stitching threads, seams, pants cuffs, pockets, under the flap of shirt pockets, and abraded areas where protruding fibers may entangle a bit of foreign debris. Older deposits of waxy, sticky, or gelatinous food materials on a garment may retain subsequent deposits of traces and debris (Figure 8.21). In addition, particulate materials such as glass shards or paint flakes may have worked their way below the surface fibers and be found in the interstices of the weave or in threads (Figure 8.22).

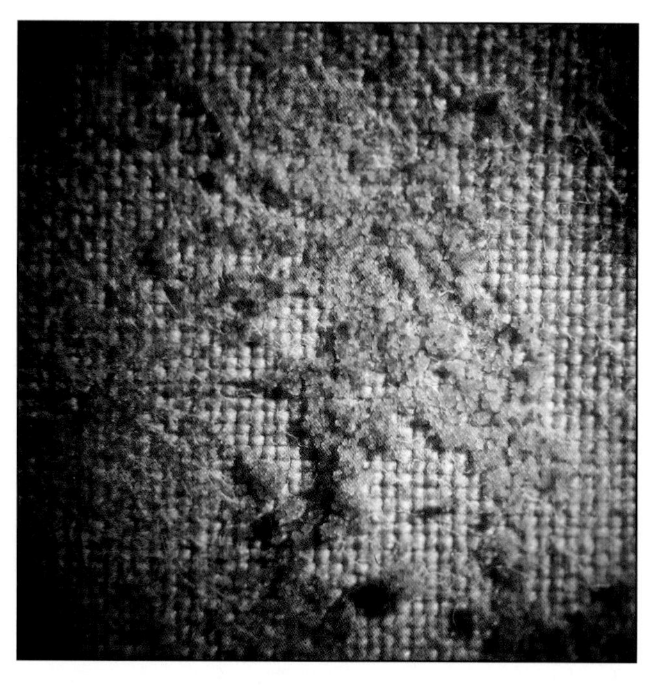

Figure 8.21 Wax drip on blue jeans with embedded fibers and particles.

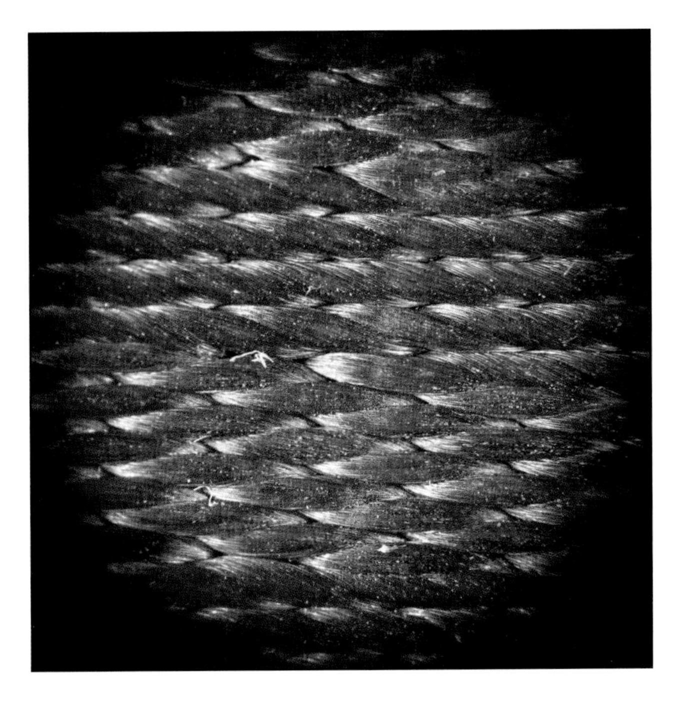

Figure 8.22 Red fibers found in the interstices of seat belt threads, partly below the surface fibers.

8.11.6 Collecting Samples for Context-Based Examinations

A context-based examination involves detailed comparison of the component fibers of two items suspected of being in contact, and a less-detailed comparison of the non-component debris; that is, the other particles and fibrous material adhering to each item. A search for component fibers is essentially a target search. Examination of the context of the deposit — the non-component debris — requires that a sample of the overall assemblage of fibers, particles, and stains be collected. A preliminary sample using a piece of clear sticky-tape allows a rapid determination of the types of samples that will be needed: a target search for component fibers and major types of non-component debris or a sampling of the overall assemblage. The latter is warranted when there appears to be a high degree of correspondence between the sets of debris on the items being compared. When this condition is not met, a modified context-based examination can be performed to help assess the significance of finding a corresponding target material.

The most efficient technique for collecting and examining an assemblage of particles is probably manual collection. Hairs should be placed into plastic bags or another suitable container and sorted later. Sampled fibers can be immediately mounted and sorted by color on a large 2″×3″ microscope slide. Particles can be sorted into groups and mounted on another slide. Such groups may include woody particles, plant parts, insect parts, round particles (pollen, clumps of starch grains, insect eggs, spores, etc.), mineral grains, possible glass shards, and opaque particles (paint chips, metal fragments, fragments of plastic, etc.). The scientist should collect samples from several areas of the garment, to control for bias in particle and fiber distribution. Dust and fine lint can be collected as a powder and sorted afterward. For discussion of follow-up testing, the reader is directed to the literature referenced in Section 8.7.

8.12 Reference Samples and Reference Standards

A reference sample represents a particular object, such as an item of clothing, or a particular person, such as a sample of a body tissue such as blood, saliva, bone marrow, etc. A reference standard represents a particular type of material (e.g., Nylon 6.6 of a certain brand and manufacturer) or a particular quantitative composition (e.g., with a certain percent of titanium dioxide delustrant) and must be of known provenance. Provenance is the supporting information that attests to the history and source of the material and is the equivalent of chain-of-custody for an evidence item.

8.12.1 Reference Samples

In addition to sampling traces and debris, the clothing examiner should also collect reference samples of the clothing fabric and reference samples of debris that might characterize the clothing and its wearer(s).

An adequate reference sample represents the composition of a material, includes all its components, reflects the range of variation, and does not include anything extraneous. Reference samples of the component fabric of a garment should include the weave and pattern and some of the stitching threads, and it must include all the types and colors of fibers and threads in the material, including fabric from thinned, discolored, or stained areas, and a sample of the stitching threads. Cut a swatch from an area that does not appear to exhibit other deposits or damage. Ensure that the swatch cut as a reference reflects the fading and abrasion exhibited by the areas of potential contact. If there is no suitable area for cutting a swatch, or if it is important for the garment not to be cut, take close-up photographs of the threads and fabric and obtain a sampling of fibers of each type using forceps.

Samples of stains, deposits, and debris should be collected as reference samples for materials that may have transferred to another item. The predominant type of fiber pills should also be collected as material that may transfer, as illustrated in Case Example 8.1, in which a fiber pill found on the victim's buttocks corresponded with fiber pilling in the crotch area of the suspect's cotton/polyester blend pants. The crotch area was worn and abraded to the extent that no cotton remained on threads or fiber pills from that area. Collect reference samples of debris from pockets and pants cuffs. Collect secondary hair reference samples from the upper back of shirts and tops.

8.12.2 Reference Standards

Reference standards of fibers, fabrics, and dyes can be obtained directly from manufacturers or purchased from vendors who procure them from manufacturers for use as reference standards. If procured directly from the manufacturer, the documentation attesting to the fabric or fiber origin must be accessible. If certain properties are of interest, the variation of the property in a fabric or fiber of that type must be established either by the manufacturer or by a reference laboratory.

Reference standards may also be needed to represent the effect of processes on a particular type of material. For example, paint manufacturers perform standard weathering tests by exposing paint reference standards to the elements in different climates and at different degrees of exposure (Hamburg and Morgans, 1979). Clothing items are subjected to tests involving laundering conditions, exposure to sunlight, and flammability. Manufacturer reference samples that reflect such testing may be difficult or impossible to obtain, and manufacturers do not test for exposure to the elements, to body decomposition fluids, or to burial. The forensic science literature includes a few studies (Ryder and Rowe, 1997), but no reference samples are currently commercially available. Such samples must be procured by experiment, preferably using reference standards of known provenance or purchased garments bearing a tag with a manufacturer's code. However, the provenance of greatest interest is documentation of the testing process and results and the preserved samples of unexposed material.

8.12.3 Primary and Secondary Reference Samples

A primary standard or reference sample is known to be from the source it represents. A secondary reference sample is reasonably presumed, but not known, to be from the source material. Obtaining secondary standards (secondary reference samples) of hair from the upper backs of shirts and tops was discussed earlier; this is usually employed if a person has been reported missing. Secondary reference samples may also be used if the source item for comparison is no longer available, as illustrated in Case Example 8.16.

When a secondary reference sample is used, it must first be examined for consistency, and any outliers must be separated and retained apart from the rest of the sample. The particles or fibrous materials with normative values should be used for comparisons. Some samples may exhibit more than one predominant value. An example is tempered glass, which may exhibit a different refractive index on the surface than in the interior. Both values would then be used in comparisons. Hair, which may exhibit several normative groups, provides another example.

CASE EXAMPLE 8.16

Secondary reference sample; glass and glass damage; avoiding contamination; preliminary sampling

A homeowner looked out the window when she heard glass breaking. She saw someone running from her shattered basement window holding, then dropping, a crowbar. Police responded and found not only the crowbar but also a pair of gloves. A man was later detained in the vicinity. The gloves, crowbar, and the suspect's hooded sweatshirt (fleece) were submitted to the laboratory, but no reference samples of glass. Reference samples were immediately requested, but the homeowner had already replaced the window and cleaned up the shards. Several shards of glass were collected from the gloves and from the sweatshirt. In addition, tiny intermittent very sharp glass cuts were observed in both garments (see Chapter 6). When the scientist examined the crowbar, no obvious glass was observed, but a white powder on the elbow of the crowbar proved to be pulverized glass. The refractive index range of the glass shards in the pulverized sample were narrow, likely to be from a single source, thus a suitable secondary reference sample. Because the homeowner saw the crowbar being dropped by the fleeing intruder, it could be presumed, although not known, to be from the broken window.

8.13 Reconstruction of Events

The following two case examples illustrate the use of traces and debris in reconstructing an event and in linking a series of incidents with a suspect.

CASE EXAMPLE 8.17

Reconstruction of events: position of shooting victim determined from fibers on bullet and bullet damage to clothing

A man detained on a city street for questioning by police died by police fire, shot in the back. The officers reported shooting a fleeing felon who had been resisting arrest. Firearms examination confirmed that the several bullets recovered at autopsy were fired from the officer's service revolver, and the crime laboratory found several bullet holes in the victim's clothing. It was winter, so the man was wearing a thick wool coat, a floral print shirt, an undershirt, pants, and a hat with a brim. One bullet did not retain fibers. Another bullet penetrated the pants. Four of the bullets exhibited fiber tufts from the wool coat and the shirts. The floral print of the shirt consisted of lighter blue and darker blue flowers. Only one bullet exhibited fibers from penetrating a dark blue flower, stacked at the top of a fiber plug that included fibers from the wool coat (two layers) and the wool felt hat worn by the deceased. The bullet defect in the dark blue flower was located approximately mid-shirt-back, thus the lower back of the body. The fibers attributable to the felt hat were found only on this bullet. The hat was damaged at the edge of the brim in the back, exhibiting abrasion and partial heat damage to the wool felt fabric. How could this have happened?

In attempting to reconstruct the shooting position that would have allowed both these areas to be in the path of the same bullet, scientists noticed a slight abrasion to the collar at the back of the neck, suggesting the collar had been partly raised and in the bullet path. This trajectory could be explained if the man's body had been arched, with his head tilted all the way back. This suggested that the man was shot while restrained with his hands behind his back, not while fleeing.

CASE EXAMPLE 8.18

Reconstruction of events: linking a series of incidents and a suspect; target search for clothing of specific colors

A series of attempted rapes occurred in three cities along a 120-mile stretch of freeway. Each victim reported that an assailant forcibly entered her vehicle and attempted to rape her. The first two vehicles were abandoned some distance from the victim. In the third case, the victim drove to the police station after a man forced his way in. He got stuck in the sunroof trying to flee and finally managed to extricate himself in the police parking lot. He fled a short distance to a busy shopping area, where he was arrested. The forensic scientist who examined the suspect's clothing for biological fluids described what appeared to be blue paint on the shoulder of his T-shirt. The paint was smeared in the location and direction that would be expected in the scenario described. The T-shirt was forwarded to another laboratory to be examined for trace evidence. The blue paint was analyzed and found to correspond with the paint of the vehicle with the sunroof, corroborating that the person arrested was the man stuck in the sunroof.

Police found out about similar incidents in the other two cities and submitted clothing of those victims. The detectives who had collected samples of debris from the driver's seat of each vehicle noted the colors of predominant debris fibers and obtained clothing items of those colors from each victim. The probable source of each major fiber type was thus determined. Fibers corresponding to major debris fibers from each vehicle were found on the suspect's T-shirt, despite his wearing it for three days. In addition, debris from the driver's seat in the second case included fibers that corresponded with the major fiber types from debris in the first case. Debris from the third case included fibers corresponding with major fiber types from the first two cases. This established a link between the suspect and a series of incidents involving a sequential transfer of debris. Several human head hairs from two of the vehicles corresponded with reference samples representing the suspect. In addition, the suspect could not be eliminated as a source of semen found in vaginal swabs from the first two women.

Because these incidents occurred before the advent of routine DNA testing, more conclusive findings were not possible from the semen in the vaginal swabs. If these cases occurred today and semen in the vaginal swabs was attributed to the suspect, especially sequential deposits, the detailed study of debris transfers would be less important to solving this case. However, let us consider what would happen if the case details changed slightly. Let us suppose that a suspect used a condom, did not ejaculate, or raped the women with an object. In these alternative scenarios, examinations of debris would be highly significant.

8.14 Process-Based Descriptive Terminology for Traces and Debris

Extruded: Elongated by passage through an aperture while viscous. Examples include fibers, hairs, pasta, animal scat, etc.

Drawn: Elongated by pulling. Examples include metal wires, fibers spooled under tension after extrusion, elongated bits of plastic, fiberglass, glass fibers drawn by falling from larger molten pieces, etc.

Molded: Hardened or set while in a mold, taking the shape of the mold. Examples include many plastics (such as those with a tell-tale raised line), concrete, natural concretions such as mudstone, fossil stone crystallized in natural structures, etc. This is sometimes observed on a micro scale.

Chambered: A chamber is a space enclosed by walls. These can either be constructed, as in beehives or wasps' nests, or produced by gas expansion. The latter can be observed in popcorn, popped rice used in certain teas, foam rubber, Styrofoam, soap bubbles, and heat-decomposed polymer deposits. Debris from foam rubber chambers may include Y-shaped fragments of the walls. Certain reflective materials, such as those used in auto license tabs, are chambered.

Hollow spheres: Spheres form when the pressure from all directions is equal. Hollow spheres result from expanding liquids or gases and are observed in conjunction with heat damage, pyrotechnics, black powder firearms, various industrial processes, and natural fires. May have "tails."

Drop-molded (beaded): Spheres formed when molten material is dropped through air or into a liquid. May have "tails." Examples include glass beads and metal beads. Beads of metal from soldering or welding may appear on clothing, sometimes accompanied by heat damage to the fabric. Glass beads may be found in paint smeared on clothing from impact with reflective lines on a roadway.

Spheroids: Formed by a variety of processes. Examples include insect eggs, globules of starch grains (the grains themselves are not resolved using a stereomicroscope), mold spores, beads placed in planting soil to hold moisture, some desiccants, some seeds, pepper corns, some mineral grains, etc.

Polished: Finely abraded, resulting in smoothing. Examples include tumbled stone, at the beach or at the rock shop.

Corroded: Oxidized metal, resulting in pitting and loss of material in the form of metal salts. These may be observed as microcrystalline deposits. Examples include rust and other oxides and, under certain conditions, carbonates. See also *Etched.*

Etched: Pitted by acid or other chemical processes, resulting in a frosted appearance in glass or removal of material in metal, sometimes producing patterns. See *Corroded.* The action of seawater etches beach glass; the glass at the edge of a microscope slide may be frosted.

Tunneled: Produced by insect, fungal, or microbial boring action involving displacement of material. May appear on bits of bark or leaf, in hair, or on a food deposit. Also includes solvent or gas tunneling, such as observed in some volcanic glasses, and rarely in fiber melt on clothing items.

Curdled: A wrinkled skin can form on a material upon rapid change of physical state. Scrambled eggs are one example.

Pressed: Deformed by pressure, producing indentations or impressions. Examples include the seal of zip-lock style plastic bags and impressions of objects that were pressed against a softer deposit.

Rubbed: Mildly abraded by contact with motion against a relatively smooth surface, resulting in transfer of particulate residues, soft films, or curled particles, as in rubbings of rubber.

Ruptured: Forcibly pulled or broken apart; the damage margins of such material.

8.15 Trace Evidence Recovery Guidelines

Detection, Collection, and Preservation Techniques

Trace evidence recovery or collection techniques used should be the most direct and least intrusive technique or techniques practical. Collection techniques include picking, lifting, scraping, vacuum sweeping, combing, and clipping.

Picking. Trace evidence may be separated from an item by using clean forceps or other implements. The collected samples should be immediately protected against loss or contamination.

Lifting. An adhesive-bearing substrate such as tape is repeatedly and firmly patted or rolled over the item causing loosely adhering trace evidence to stick to the tape. Do not overload the tape. The collected lifts are typically placed on a transparent backing (e.g., clear plastic sheeting, glass slides, and clear plastic or glass petri dishes). This protects against contamination and permits samples to be easily viewed and removed for further comparison.

Scraping. A clean spatula or similar tool is used to dislodge trace evidence from an item onto a collection surface such as clean paper. The collected debris is immediately packaged in a manner to avoid sample loss. This technique is most often conducted within the laboratory in a controlled environment that reduces the risk of contamination or loss of the trace evidence.

Vacuum sweeping. A vacuum cleaner equipped with a filter trap is used to recover trace evidence from an item or area. The filter and its contents should be immediately packaged to avoid sample loss. The appropriate vacuum parts, filter, and trap must be changed and rigorously cleaned between each vacuuming to avoid contamination. Consider using this method subsequent to other collection techniques as it is indiscriminate and may result in the collection of a large amount of extraneous material.

Combing. A clean comb or brush is used to recover trace evidence from the hair of an individual. The combing device and collected debris from the hair should be packaged together.

Clipping. Trace evidence can be recovered from fingernails by nail clipping, scraping, or both. Fingernails may be clipped with clean scissors or clippers and packaged in clean paper. Fingernails may be scraped with a clean implement to collect debris from under the fingernails. Package the collected debris and the scraping device as one unit, typically in clean paper. Commonly, fingernails from the right and left hands are packaged separately. This does not preclude the collection of each or any nail separately from all others, such as a nail with obvious damage.

8.16 References

Bisbing, R.E., *Forensic Hair Comparisons: Guidelines, Standards, Protocols, Quality Assurance and Enforcement*, Presentation to the National Academies Committee on Identifying the Needs of the Forensic Sciences Community, April 24, 2007.

Blackledge, R.D., Ed., *Forensic Analysis on the Cutting Edge: New Methods for Trace Evidence Analysis*, Wiley-Interscience, Hoboken, NJ, 2007.

Chable, J., Roux, C., and Lennard, C., Collection of fiber evidence using water-soluble cellophane tape, *J. Forensic Sci.*, 39, 1520–1527, 1994.

Chisum, W.J. and Turvey, B.E., *Crime Reconstruction*, Elsevier Academic Press, Burlington, MA, 2006.

Cook, R., Evett, I.W., Jackson, G., and Jones, P.I., A hierarchy of propositions: Deciding which level to address in casework, *Sci. Justice*, 38(4), 231–239, 1998.

Coxon, A., Grieve, M.C., and Dunlop, J., A method of assessing the fibre shedding potential of fabrics, *J. Forensic Sci. Soc.*, 32, 151–158, 1992.

Crutcher, E.R., *Assemblage Analysis — Identification of Contamination Sources*, 2nd Spacecraft Contamination Conference, 1978.

Cwiklik C., *A Case Study: Mt. St. Helens Volcanic Ash*, Presented to the American Academy of Forensic Sciences, Los Angeles, CA, February 1981.

Cwiklik, C., *Particle Sizing: Sieving as a Microanalysis Method*, Presented to the Northwest Association of Forensic Scientists, Seattle, WA, 1982.

Cwiklik, C., An evaluation of the significance of transfers of debris: Criteria for association and exclusion, *J. Forensic Sci.*, 44(6), 1136–1150, 1999.

Cwiklik, C., *Questions of Inference: I. Microscopic Grouping of Hairs and Selection of Hairs for DNA Testing. II. Transfer of Hair in the Context of Groupings of Hair and Sets of Debris*, AAFS meeting, Seattle, WA, February 2001.

Cwiklik, C., Hair microscopy in the age of DNA testing, *International Association of Microanalysts (IAMA) newsletter* (http://www.iamaweb.com), January 2003.

Cwiklik, C. and Dean, M.J., *Heat Damage to Cotton Fabrics as a Clue to the Conditions That Produced It*, Presented to the American Academy of Forensic Sciences (AAFS), Washington, DC, February 2008.

Cwiklik, C.L. and Gleim, K.G., Forensic casework from start to finish, in *Forensic Science Handbook*, Vol. III, pp. 1–30, Saferstein, R., Ed., Prentice-Hall, Englewood Cliffs, NJ, 2009.

Deadman, H.A., Fiber evidence and the Wayne Williams trial. Part I., *FBI Law Enforcement Bull.*, 53(3), 12–20, 1984a.

Deadman, H.A., Fiber evidence and the Wayne Williams trial. Part II, *FBI Law Enforcement Bull.*, 53(5), 10–19, 1984b.

Grieve, M.C. and Biermann, T.W., Wool fibers — Transfer to vinyl and leather vehicle seats and some observations on their secondary transfer, *Sci. Justice*, 37, 31–38, 1997.

Griffin, H., Glass cuts, in *Forensic Analysis on the Cutting Edge: New Methods for Trace Evidence Analysis*, Blackledge, R.D., Ed., Wiley-Interscience, Hoboken, NJ, 2007.

Hamburg, H.R. and Morgans, W.M., Eds., *Hess's Paint Film Defects — Their Causes and Cure*, Chapman and Hall, London, 1979.

Horrocks, M., Coulson, S.A., and Walsh, K.A.J., Forensic palynology: Variation in the pollen content of soil surface samples, *J. Forensic Sci.*, 43(2), 320–323, 1998.

Kosanke K.L., Dujay R.C., and Kosanke, B., Pyrotechnic reaction residue particle analysis, *J. Forensic Sci.*, 51(2), 296–302, 2006.

Linch, C.A. and Prahlow, J.A., Postmortem microscopic changes observed at the human head hair proximal end, *J. Forensic Sci.*, 46, 15–20, 2001.

Locard, E., The analysis of dust traces, *Am. J. Police Sci.*, 1930.

Lowrie, C.N. and Jackson, G., Secondary transfer of fibres, *Forensic Sci. Int.*, 64, 73–82, 1994.

McCrone, W.C., Particle analysis in the crime laboratory, in *The Particle Atlas*, McCrone, W.C., Delly, J.G., and Palenik, S.J., Eds., Vol. V, pp. 1379–1401, Ann Arbor Science Publishers, Ann Arbor, MI, 1979.

McQuillian, J., A survey of the distribution of glass on clothing, *J. Forensic Sci. Soc.*, 32, 333–348, 1992.

Ogle, R.R. and Fox, M.J., Atlas of human hair microscopic characteristic variates, *Sci. Justice*, 38, 55, 1998.

Palenik, S.J., Microscopic trace evidence — The overlooked clue: Part I, *The Microscope*, 30(2), 93–100, 1982a.

Palenik, S.J., Microscopic trace evidence — The overlooked clue: Part II, *The Microscope*, 30(3), 163–169, 1982b.

Palenik, S.J., Microscopic trace evidence — The overlooked clue: Part III, *The Microscope*, 30(4), 281–290, 1982c.

Palenik, S.J., Microscopic trace evidence — The overlooked clue: Part IV, *The Microscope*, 31(1), 1–14, 1983.

Parybyk, A.E. and Lokan, R.J., A study of the numerical distribution of fibres transferred from blended fabrics, *J. Forensic Sci. Soc.*, 26(1), 61–68, 1986.

Petraco, N., Trace evidence — The invisible witness, *J. Forensic Sci.*, 31(1), 321–328, 1986.

Petraco, N. and Fraas, C., Morphology and evidential significance of human hair roots, *J. Forensic Sci.*, 33, 68–76, 1988.

Robertson, J., *Forensic Examination of Human Hair*, Taylor & Francis, London, 1999.

Robertson, J. and Roux, C., Transfer, persistence and recovery of fibers, in *Forensic Examination of Fibers*, Robertson, J. and Grieve, M., Eds., Taylor & Francis, London, 1999.

Ryder, M.L. and Rowe, W.F., Biodegradation of hairs and fibers, in *Forensic Taphonomy: The Postmortem Fate of Human Remains*, Haglund, W. and Sorg, M.H., Eds., CRC Press, Boca Raton, FL, 1997.

Salter, M.T., Cook, R., and Jackson A.R., Differential shedding from blended fabrics, *Forensic Sci. Int.*, 33, 155–164, 1984.

Scientific Working Group on Materials Analysis (SWGMAT) Evidence Committee, Trace evidence recovery guidelines, January 1998 revision, *Forensic Sci. Commun.*, 1(3), October 1999.

Suzanski, T.W., Dog hair comparison: Pure breeds, mixed breeds, multiple questioned hairs, *Can. Soc. Forensic Sci. J.*, 22(4), 299–309, 1989.

Taupin, J.M., Hair and fiber transfer in an abduction case — Evidence from different levels of trace evidence transfer, *J. Forensic Sci.*, 41(4), 697–699, 1996.

Taupin, J.M., Forensic hair morphology comparison — A dying art or junk science? *Sci. Justice*, 44(2), 95–100, 2004.

Teetsov, A.S., An organized approach to isolating and mounting small particles for polarized light microscopy, *Microscope*, 50(4), 159–168, 2002.

Thorwald, J., *Crime and Science*. Harcourt, Brace and World Inc., New York, 1967.

Wiggins, K., Emes, A., and Brackley, L.H., The transfer and persistence of small fragments of polyurethane foam onto clothing, *Sci. Justice*, 42(2), 105–110, 2002.

Results and Their Significance 9

9.1 Significance of the Evidence

The purpose of forensic clothing examination is to provide information to investigators and attorneys and evidence in a legal case. The clothing examiner must provide the justice system and the public with good science: work that the public, the law enforcement and legal communities, the courts, and individuals accused of crimes can be confident is useful, complete, and fair. This implies work that is rigorous, thorough, and germane to the overall issues in the case. The forensic scientist must translate the overall case questions of the investigators and the attorneys into those that science can address, asking "What can we really answer?" The examiner cannot provide answers as to the guilt or innocence of a person but can consider questions relating to the "who, what, where, and how" of a case, and can even provide evidence as to whether a crime has been committed, as the following example shows (Taupin, 2000):

> A middle-aged Melbourne woman claimed she had been raped by an unknown male while walking through bushland. She alleged that he grabbed her top and ripped it open horizontally, ripped her brassiere with a knife and then used both his hands to rip the fly of her jeans open; he subsequently inserted an object into her vagina. The examination of the woman's clothing revealed scissor cut type damage to her top and brassiere and cuts to the stitching of the fly of the jeans. When confronted with this evidence the woman admitted she had concocted her story in order to obtain money from the Crimes Compensation Tribunal. She was subsequently charged with false report.

This case example underscores the importance of providing investigative information in the early stages of a case, even before a suspect is accused of a crime and well before a defendant is charged and scheduled for trial. The finding of evidence must be timely in order to be of the utmost value in the investigation stage of a case. Police understand that most crimes are solved within a few days of their commission. Forensic scientists consequently find constant demands to come up with evidence and answers in a very short time frame. Crime shows on television increase the expectations of investigators. The almost instantaneous provision of forensic evidence to detectives in these shows makes one wonder why "backlogs" exist in laboratories. Because it is not usually possible to perform rigorous, thorough, and complete scientific work on complex cases within only a short time, it is incumbent on the examiner and the laboratory to establish impact-based priority systems that allow for rapid work to be performed on the most urgent and useful examinations, and for the remaining work to be performed later (Cook et al., 1998; Cwiklik and Gleim, 2009).

The search for evidence and its interpretation should not be compromised by strict time frames or rushing through the process in order to obtain a result. The evidence must be of high quality, and any rush to obtain it may compromise this quality. The clothing examiner needs to work with the investigator in establishing priorities so that time is spent on the most urgent and useful examinations, which can be provided at the outset only if the remaining work is postponed.

Although seldom the most urgent or initially useful, the issues that can be addressed by reconstruction may be among those that determine whether the initial information is significant. The courts determine the ultimate significance of evidence, including the clothing examination, in a legal case. The scientist is seldom aware of all the other evidence involved in the case and thus will not always be aware of the relative value of the clothing evidence with respect to other evidence. In reporting the results of the clothing examination, the examiner should express the implications of the results on those case questions of which he or she is aware.

It is helpful for the report to include a summary of the case information received by the examiner. Not only does this present a clear reference point for the examiner's opinion, but it also allows any errors or misconceptions in the background information to be corrected. This is especially valuable in controlling for biases introduced by those who provided the information and renders the process transparent. It allows for the expression of the examiner's opinion while maintaining the objectivity of the testing process and testing decisions. A scientific opinion should be a considered judgment. When reaching an opinion, the examiner applies the results and the scientific conclusions to the case issues expressed via background information, reaching inferences as to how the conclusions address the questions of fact. This allows the courts to determine any *legal* significance.

Scientific opinion or inference must rest upon a basis of accuracy and reliability. The significance of quantitative data such as DNA profiles or other data derived from chemical instrumentation can be evaluated with statistical tools. Other measures of significance must be used for non-numerical data. Although numerical confidence intervals cannot be used with qualitative data, the scientist needs to know the limitations of the tests and of the results. The limitations on a test are a measure of scientific accuracy and are an essential component of the significance of the findings.

The examiner should consider the following in reaching inferences and forming an opinion:

1. Access: the suspect population is confined to those with access to the crime scene
2. Transfer: the suspect, victim, or crime scene must shed and transfer the pertinent evidence in sufficient amount and condition to interpret the findings
3. Discrimination: the estimated frequency of the evidence in the relevant population
4. Context: sources of secondary transfer and indirect association
5. Concordance: the fit of results or conclusions

9.2 Expectations

A reasonable expectation on the part of investigators, attorneys, and the courts is that the information gathered from clothing examination is relevant, timely, and of high quality. Both science and the law address relevance, and both consider those facts and statements relevant that address the case issues. The scientist must be able to detect the physical evidence as well as recognize its relevance to the case. The clothing examiner, in particular, must observe and document a plethora of details that may or may not be pertinent to the case and, as the examination proceeds and the case develops, determine which of the stains, deposits, types of damage, or even the overall condition of the garment merits follow-up testing and analysis.

9.3 Context of Evidence Obtained from Clothing

The processes integral to the practice of forensic science are identification, association through class and individualizing characteristics, and reconstruction (Inman and Rudin, 2001). All these processes may be used in the examination of clothing and the interpretation of evidence collected or observed.

The clothing examiner should play a role similar to a crime reconstructionist. Access to crime scene reports and other expert reports such as autopsy findings may be necessary to place the clothing evidence in the context of the case. The history of the garment is also important in distinguishing between normal use (and other recent use) and the effects of any participation in the incident under investigation. The scientific examination is carried out within a "framework of circumstances" (Evett et al., 2000), which includes the history of the exhibit and the context of the evidence. Information about times and actions, for example, will inform judgments about transfer and persistence in trace evidence materials found on clothing, including trace DNA. The clothing examiner should explain how this framework appeared at the time of the examination. The laboratory bench notes should include a record of the background information provided, whether in conversation or correspondence or from copies of reports. It is also advisable to give a signal that, should the framework change in any way, it will be necessary to review the interpretation. This purpose is served by a summary of background information in the report, as well as an explication of any assumptions used in reaching inferences.

The issue of background information is a contentious one, with concerns that such information can create a conscious or subconscious expectation regarding the results. These expectations are often discussed in terms of "observer effects" or "context effects" (Saks et al., 2003). A comprehensive discussion of bias and controls for bias is beyond the scope of this book, except to mention that the scientific method was designed as a control for bias. Many of the publicized instances in which examiner bias influenced results of analysis appear to be instances in which the scientific method was not applied. The practice of science involves hypothesis formation and testing and the weighing of alternative hypotheses. Hypotheses cannot be formed in the absence of information about the issues that are the subject of inquiry. It would be better to structure the formation of multiple hypotheses at the outset (Platt, 1964) into laboratory practices and protocols than to cede hypothesis formation about physical evidence to investigators and lawyers, neither of whom are charged with being objective or disinterested in the outcome of a case. The scientist should be disinterested in the advocacy of either side of the adversarial system and should decline to accept work should a case arise in which this is not possible.

Nevertheless, the clothing examiner should consider the criticism that forensic science practice is far behind most scientific fields in controlling for observer effects (Saks et al., 2003). An analytical plan cannot be developed in the absence of background information, nor can an intelligent examination of a crime scene or an item of clothing, nor a sampling plan. It appears that controls on bias are best applied during the sample testing stage. The original examiner can solicit a second opinion, on the test data only, by an examiner who does not know the background information of the case or even the other test results.

9.4 Objectivity and Opinion

Objectivity is the foundation of scientific method based on facts and observations; findings should be reproducible. Objectivity is fostered by a focus on data. That is why an objective description of stains, deposits, and damage has been emphasized in this book. Objective data underlies valid results and conclusions. If the clothing items have been handled by another examiner and if samples have already been collected, the present examiner should be careful to account for any biasing of his or her observations. A simple example is the circling of stains by the first examiner. The circles around some stains may affect the pattern perception of the second examiner and make it more difficult to notice other stains and deposits. One way to counter this would be to trace the stains, but not the circles, onto a piece of clear plastic, then examine the tracings for any patterns.

The role of experience is also important in clothing examination and other scientific endeavors. Experience is by its nature personal and subjective. It is at the heart of the "I wonder if" that leads to discovery. Case Example 4.1 describes a case in which a splash of bean soup led to an insight into a possible murder site. The proper role of experience is to frame questions that can be tested objectively. It should not be the basis of conclusions. The examiner's experience-based intuition suggesting bean soup was followed by cooking and testing a variety of beans. Forming inferences or arriving at opinions, albeit based on objective findings, includes an element of subjectivity arising from the examiner's experience and assumptions. This is legitimate provided that the subjective element can be articulated. As long as the opinion has been formed through hypotheses testing and other checks and balances as described in this book, then it is a meaningful and valid one. Although many philosophers maintain that strict objectivity is an unattainable goal and that all views are to some extent influenced by the observer, it is incumbent upon the scientist to identify those elements to the best of his or her ability and to be open to considering other possibilities.

The scientific method is predicated on objective observations and reasoning. The scientist can be expected to be objective, disinterested, and as unbiased as possible:

- Being objective means basing conclusions on facts, not on unstated assumptions or a preconception about an outcome.
- Being disinterested means that one is impartial and has no stake in the outcome.
- There are many possible stakes, including recognition, advancement, ego, satisfaction, and so on, but these are not insurmountable, and the classical tools of the scientific method act as controls.

Bias arises partly from the information at hand (Nordby, 1992). It can only be minimized, not completely eliminated. The scientist may know some case particulars but not others. He or she has a particular experience and knowledge; that is, an expertise. Another source of bias is the way the brain works — the biologically programmed strategies for knowing.

The scientist can control for the first type of bias by making the background knowledge and working assumptions explicit. If the case information available to the scientist is restricted so that the scientist works "blind," then examiner bias can be reduced, but the effect on laboratory work of any police bias or misinformation is increased. This subverts the function of laboratory work in testing and questioning the police hypotheses and in providing objective and independent information that might not have been anticipated by police. Another effect would be to deprive police of investigative information that the laboratory can provide (Cwiklik, 2006).

The second type of bias, arising from one's experience and knowledge, is readily controlled for by a collegial team approach. For example, a botanist tends to place more significance on plants than on minerals, even if the minerals are equally important as evidence, because to a botanist plants are a richer source of information. Sometimes the "team" includes colleagues from other laboratories who can provide useful discussion.

The third source of bias is the way the brain works. When the brain's usually effective strategies are misused or overused, errors in judgment can result (Gilovich, 1991; Kahneman et al., 1982). There is an accessible body of literature dealing with types of errors in judgment and reasoning. Many scientists strive to be open-minded. An awareness of common cognitive errors can be helpful to scientists in devising ways to control for the biases that may interfere.

9.5 The Adversarial System and the Law

Lawyers are trained to approach truth through argument, whereas scientists are trained to approach truth through experiment (Schwartz, 1981). The scientist uses argument in designing a testing plan and troubleshooting a theory, and the lawyer uses facts and test results as points of argument.

The scientist is often more comfortable at the investigative stage of a case. This is the stage of finding things out, of following hunches and testing them. When a case is in the trial stage, the scientist is still engaged in testing and narrowing down hypotheses as the case focuses on the defendant. However, if the case is first presented to the scientist at this stage, the stage of open inquiry has been bypassed. The scientist understands that no case is perfect and evidence is never complete and will instead tend to weigh the strengths and weaknesses of the evidence against the defendant rather than asking if the direction of inquiry was appropriate. This is a bias toward the information at hand in contrast with information that has not been developed. The scientist can control for this by forming more open hypotheses as to what actions may have occurred, despite the narrower focus of the legal stage of the case.

The scientist is trained to be objective and to consider alternative explanations equally, but the adversarial legal system in many countries is constructed around advocacy. This makes for a dichotomy between science and the law. However, it is the law that asks the relevant questions (e.g., "Did this suspect commit this crime?"), and it is the court that determines the relevance and admissibility of the evidence. The forensic scientist cannot answer the ultimate question and should form no opinion as to guilt or innocence, which are distinct from actions taken or not taken. What the scientist *can* do is search for relevant evidence, and then interpret that evidence, whether that evidence is inculpatory or exculpatory.

9.6 Interpretation and Communication of the Evidence

In addition to providing the results of the examination and subsequent testing of the results in report form and testimony, the clothing examiner should place these results in the context of the case. There should be a conclusion as to the overall meaning of the results, including how results of various types of testing fit together. In the conclusion, the examiner explores all reasonable interpretations of the data. The interpretation of the observations is the heart of forensic science.

Any differences between the evidence and reference samples are explained as either actual differences inherent in the evidence or as a consequence of the imprecision of the test system. The limitations of the test and the evidence, the conclusions, the assumptions, and the inferences that follow from them are enunciated. The results of all examinations are incorporated in the report, including results that at the time the report is issued may seem irrelevant. Seemingly irrelevant information may be of importance if a new suspect is developed, a different scenario is considered, evidence tampering and misdirection are revealed, or other issues develop that are not known to the scientist. One of the authors was asked to examine a seat belt for any evidence that the passenger was wearing it at the time of an impact injury. A large clump of dust and debris fell out of the seat belt mechanism onto the butcher paper the seat belt was placed on and was mentioned in the report only in the section on evidence packaging. This was raised during testimony, because one of the issues in the case, not known to that scientist, was that some seat belts fail due to deposits that interfere with the functioning of a small metal tongue in the mechanism. The scientist is on firm ground by issuing a complete report. Finally, an interpretation is without meaning unless the clothing examiner clearly states the alternatives considered. Less likely but possible alternatives should be reported as such and unsupported alternatives explained.

The results of the clothing examination should be communicated as simply as possible and using uncomplicated language or terminology. Clothing items are familiar to the public and the judiciary because they are worn by them, chosen by them, laundered by them, and even fashioned by them. Clothing examiners should use this to their advantage and phrase their results, either in report form or in testimony, in a way that can be readily understood by the public.

9.7 Peer, Technical, and Administrative Reviews

Both the laboratory and the individual scientist employ systems that recognize that scientists are human beings and can thus make errors and that both people and computerized instruments have biases in decision-making. Some of the systems are designed to prevent errors and others to identify and resolve them if they nevertheless occur, so that the results and conclusions issued by the laboratory are sound. Furthermore, a laboratory typically employs systems to ensure work of adequate quality. In a healthy laboratory, open discussion of errors and their prevention is encouraged.

Peer review — having a fellow-scientist review one's work — can be one of the cornerstones of ensuring work of high quality. In-house peer review is of great value when performed thoughtfully and less so if performed simply to ensure compliance with protocols. Peer review by someone outside of the laboratory is usually performed when work is reviewed by a scientist retained by the opposing side in a legal case. This type of review is more likely to be free of institutional biases but often suffers from lack of direct communication between the scientists — often restricted by the attorneys.

Peer review may consist of a scientist reviewing a colleague's scientific report and data, as found in the laboratory bench notes, or may involve checking results and conclusions. It may involve a second opinion or a discussion of the analytical plan, the conclusions, the inferences, or any other decision-making step. The scientist may request a blind second opinion on results or an informed discussion of a reconstruction. The basic tenet of quality assurance is that significant decisions are checked by a colleague. When done well, peer review and the attendant critical discussion of the work on a case can stimulate the first scientist in his or her thinking and can improve not only the work on a particular case but the work of the scientific group as a whole.

The object of direct peer review of a result is to check that the test has been performed according to quality guidelines, not to belittle the work of any individual. For example, the rationale behind a DNA test of a bloodstain or the results of a dye test of a particular fiber should be noted by the forensic caseworker performing the test and then checked by a colleague.

The report itself should also be subject to thorough review. A technical or peer review of a report is the independent review of the report by another scientist in the relevant discipline, followed by a discussion of the salient points. Furthermore, these discussions promote a sharing of experiences that can only result in the improvement of quality in the laboratory. An administrative review ensures that the report can be understood by either a lay reader, such as an investigator, attorney, or member of the jury, or by another scientist, including a scientist outside one's immediate discipline. These reviews are typically required by the standard operating procedures of an accredited laboratory.

The rationale behind a second opinion is to provide a fresh perspective on a difficult analysis and to provide a cross-check for any blind spots of the initial examiner. A second opinion is more valuable if the second examiner takes notes on the examination and places the notes into the case file. This prevents confusion as to exactly what is being reviewed and in which observation, result, or conclusion the peer review supports the first scientist. When disagreements or differences of opinion arise, it is important to resolve them methodically (Cwiklik and Gleim, 2009).

The laboratory should have a system in place to resolve discrepancies between examiners. Differences of opinion may occasionally arise in scientific analyses. Ethical and moral differences of opinion should be resolved amicably and impartially according to national, accreditation, and laboratory guidelines (Barnett, 2001).

9.8 Training and Maintaining the Expertise of the Clothing Examiner

Every clothing examiner must maintain his or her expertise through training programs, attendance at seminars and lectures, regular reviews of the scientific literature, and discussions with colleagues. In the past this has been somewhat informal, but with the introduction of accreditation of laboratories, the maintenance of expertise has been formalized in quality documents of the laboratory.

Proficiency testing of the forensic examiner is now mandatory in accredited laboratories. However, the number and type of each proficiency test may be somewhat problematic for a forensic clothing examiner, because the most critical tasks in clothing examination do not usually fall into the narrower specialty areas for which commercially available proficiency tests are designed. There are difficulties in designing and evaluating even the specialty proficiency tests. Proficiency tests other than informal in-house tests have yet to be developed for clothing examinations. A suitable test would focus on recognizing and interpreting stains, deposits, and damage as well as selecting and collecting samples for further testing. This is a subject for future development. In the meantime, it is incumbent on the laboratory to ensure that every clothing examiner maintains his or her expertise to ensure that each and every result and conclusion reported is valid and reliable and each inference is justifiable.

9.9 References

Barnett, P.D., *Ethics in Forensic Science: Professional Standards for the Practice of Criminalistics*, CRC Press, Boca Raton, FL, 2001.

Cook, R., Evett, I.W., Jackson, G., et al., A model for case assessment and interpretation, *Sci. Justice*, 38, 151–156, 1998.

Cwiklik, C., *A Good Look at Blind Testing: Quality Assurance Systems and Bias*, American Academy of Forensic Sciences meeting, Seattle, WA, February 2006.

Cwiklik, C.L. and Gleim, K.G., Forensic casework from start to finish, in *Forensic Science Handbook*, Vol. III, pp. 1–30, Saferstein, R., Ed., Prentice-Hall, Englewood Cliffs, NJ, 2009.

Evett, I.W., Jackson, G., Lambert, J.A., and McCrossan, S., The impact of the principles of evidence interpretation on the structure and content of statements, *Sci. Justice*, 40(4), 233–239, 2000.

Gilovich, T., *How We Know What Isn't So: The Fallibility of Human Reason in Everyday Life*, The Free Press, a division of Simon & Schuster Inc., New York, 1991.

Inman, K. and Rudin, N., *Principles and Practice of Criminalistics: The Profession of Forensic Science*, CRC Press, Boca Raton, FL, 2001.

Kahneman, D., Slovic, P., and Tversky, A., *Judgment under Uncertainty: Heuristics and Biases*, Cambridge University Press, New York, 1982.

Nordby, J., "Can we believe what we see if we see what we believe — Expert disagreement," *J. Forensic Sci.*, 37(4), 1115–1124, 1992.

Platt, J.R., Strong inference, *Science*, 146, 347–353, 1964.

Saks, M.J., Risinger D., Rosenthal R., and Thompson W.C., Context effects in forensic science: A review and application of the science of science to crime laboratory practice in the United States, *Sci. Justice*, 43(2), 77–90, 2003.

Schwartz, R.L., Teaching physicians and lawyers to understand each other, *J. Legal Med.*, 2(2), 131–149, 1981.

Taupin, J.M., Clothing damage analysis and the phenomenon of the false sexual assault, *J. Forensic Sci.*, 45(3), 568–572, 2000.

Appendix 1

Lab Notes: A Checklist of What to Record in Notes of a Clothing Examination

Document in laboratory notes the following:

1. Internal transfers and shared custody (movement of the item between examiners at the same laboratory)
2. Packaging and labeling
   a. Agency identifiers, including case and item numbers
   b. Type of closure (taped, stapled, etc.) and whether sealed
   c. The number of seals (has it been opened and resealed? dates?)
   d. Condition of seals if seals are not intact
   e. Date that you opened a package
   f. Date that you repackaged and sealed an item
   g. Any repairs you made to the original packaging
   h. Whether you repackaged any items in new containers

3. Evidence handling
   a. What happened to a clothing item once removed from the package
      Examination surface (e.g., clean brown paper on top of cleaned bench)
      Was the clothing left to dry? (method, location, and date)
      Any items associated with the clothing? (e.g., money or knives in pockets)
      Was the clothing turned inside out or were knots disassembled?
   b. Any changes to condition of the clothing during examination
      Swabbing or mapping of stains
      Removal of evidential material, how and where that was done
      Addition to item, such as wrapping item in paper
      Moistening/wetting a deposit or particular area of the item
   c. Packaging of samples and objects removed from an item
      Samples collected
      Pocket contents
      Debris collected from clothing during examination

4. General description of items
   a. Type of garment and where worn on body (draw a diagram)
   b. Color, size, brand, style, etc.
   c. Clothing received inside out, buttoned up, etc.
   d. General description of fabric or other material (e.g. knit, blue jean denim)

e. Types of closures (buttons, zippers, velcro, etc. — open or closed?)
f. General condition (e.g., "Clothing fairly clean with few deposits, minor body odor")
    Whether well-worn or apparently new
    Dry, damp, or wet

5. Description of stains, deposits, and damage
    a. Location and appearance (note on diagram)
    b. Information from preliminary microscopic examination
    c. Observations of evidence in more detail (e.g., diagram of a cut or description of complex stain)

6. Absence of stains or deposits
    a. If no stains or deposits were observed
    b. If no stains of expected types were observed (e.g., "Numerous stains but no blood observed on the front of the pants")
    c. If stains and deposits were not found *where* they might be expected (e.g., "Fresh abrasion was observed on the right shoulder, but no blood was observed in the corresponding area inside the garment")

7. Conditions of observation
    a. Type of lighting (daylight, fluorescent light, tungsten light, alternate light source)
    b. Angle of lighting (oblique, directly above)
    c. Unaided eye or magnification
    d. For microscopic exam, the type of microscope and the magnifications

8. Levels of information
    a. *Data*: Observations made, tests used, and any numerical measurements obtained; include peer scientific check
    b. *Results*: for example, testing results
    c. *Conclusions*: May be preliminary and informal; used to make decisions regarding further examination

9. Samples collected
    a. Which samples? (material and location; *note this on diagram*)
    b. Types of samples (e.g., control samples, reference samples, standards, etc.)
    c. Type of sampling: Representative, random, range of variation, area sampling
    d. Sampling rationale [e.g. "Blood spatter appears to be from several directions; samples taken from each grouping. Drips also sampled (could be from assailant???)"]

Appendix 2

The Stereomicroscope

A stereomicroscope of good quality should be readily available for any clothing examination. A stereomicroscope is one of the most cost-effective pieces of equipment a laboratory can buy. Every laboratory should have one available for each clothing examiner. The scientist should not have to leave the immediate examination area to use one.

A stereomicroscope provides low magnifications with an erect image and full perception of depth. Stereomicroscopes normally have a long working distance and considerable depth of field and thus facilitate manipulation of the garment. Optical quality in a stereomicroscope can be defined by the resolving power and magnification of the objective lenses. The overall magnification is the product of the objective magnification and that of the eyepiece.

The objective lens is the one immediately above the sample area and forms the image. A minimum range of 10× to 30× magnification is required but not optimal. A lower magnification may be desirable, especially when taking photographs. Unless the objective lenses have very high resolving power, a higher upper magnification is required for trace evidence examinations and examinations of cut and tear damage, which would logically be conducted using the same equipment. The upper total range of magnifications should be 40× or 50× when 10× eyepieces are being used. Eyepieces of higher magnifications may be used for minimizing eyestrain, but a better solution is to use eyepieces with a wider field of view. Higher magnification eyepieces do not provide the increased resolving power that would be obtained with a higher magnification objective lens, and they have the disadvantage of reducing the field of view so that scanning an area of interest takes more time. A certain amount of personal preference will govern the decisions between scanning area and larger image, but resolving power is not a matter of personal preference.

Resolving power in a microscope is the minimum distance at which two adjacent particles can be distinguished as separate structures; for example, as two spheres rather than as particles fused into a figure 8. Various tests of resolving power are described in microscopy texts, but when comparing microscopes, a good approximation can be obtained by taking a business card and looking for the minute ink traces at the edges of the writing that are very hard to see. The ability to resolve these hard-to-see structures will permit an assessment of resolving power. In doing this, it is important to compare the images at the same objective magnification and the same eyepiece magnification and to use comparable lighting. Even the best lenses will not provide a good image without enough light to see properly.

Several types of light sources are available, including fluorescent ring lights, flexible fiber optic light bundles, and free-standing tungsten side lights. There is also a type of stereomicroscope in which the light travels through the optics rather than being reflected back from the sample (Greenough illumination), minimizing scatter. However, this is rarely the method of choice for clothing examination. Locating stains, deposits, and damage more effectively means that the technique of oblique illumination is usually employed, in which light is incident from an oblique angle. This works largely by increasing scatter, which increases the contrast between a deposit and its surroundings, making it easier to find.

For the same reason, many examiners conducting clothing examination prefer side lighting to ring lighting, because the even illumination of ring lighting can make it more difficult to find deposits. Regarding fiber optic illumination versus free-standing tungsten lamps for clothing examination, personal preference is probably the deciding factor. Fiber optic illumination permits a narrow and intense focus but requires frequent adjustment. Tungsten side lamps offer less flexibility with varying intensity and focus but require less adjustment once in place.

A rolling stereomicroscope such as a surgical microscope is an asset because one can work more effectively with larger items. The electronic controls of a modern surgical microscope are rarely needed for clothing examination and thus do not usually have to be purchased. When selecting a rolling stereomicroscope, one should check the stability of the image at the higher end of the magnification range; if the apparatus tips or vibrates too much, the image will not be stable. Similar considerations apply to a microscope on a boom stand — test it with the boom fully extended — with the additional caveat that it should be easy to adjust the height of the microscope on the boom stand during an examination.

When funds are severely limited, some compromises can be made on convenience, but optical quality should not be sacrificed. It is possible to find stereomicroscopes with good quality glass optics among those marketed for field use or for students. The overall optical quality of a brand may be uneven, but some individual microscopes will be equipped with good lenses. In these situations, the individual stereomicroscope should be selected for resolving power. In addition, at least one stereomicroscope designed for the student market uses computer imaging that is quite satisfactory.

References

Bradbury, S. and Bracegirdle, B., Introduction to light microscopy, Royal Microscopical Society, *Microscopy Handbooks*, 42, Bios Scientific Publishers, Oxford, UK, 1998.

Appendix 3

Establishing a Reference Collection

The best way to learn how to recognize common materials and how they are deposited is to examine exemplars. These can be prepared easily and can be kept in the laboratory as a reference collection. In addition, the examiner will encounter materials in daily life, including materials in their context, and will see materials being deposited or altered. These can be brought to the laboratory and studied. Common types of materials include:

- Dried body fluids, including blood, semen, nasal mucus, saliva, and urine
- Animal scat, insect parts, fish scales, and other animal detritus
- Human and animal hairs and fur; natural and synthetic fibers; pilled fibers
- Paper and cloth fragments
- Plant parts, such as seeds, husks, stalks, roots, and bark
- Raw and cooked food; cooking ingredients, including spices
- Wood chips, sawdust, and wood shavings
- Writing implements
- Beverages
- Cooking oils and greases
- Automotive oils and greases; other automotive products
- Glues and mastics
- Building materials, such as shards of glass, fragments of brick, grout, cement, sealing compound, window caulking, plaster, insulation, etc.
- Coatings and plastics, including paint, nail polish, and plastic coatings
- Soil, sand, and clay
- Powders, including flour, cornstarch, baby powder, and fine polishing abrasives
- Cosmetics
- Laundry and personal care products, such as detergents, soap, and shaving cream
- Tobacco and marijuana from smoking materials
- Firearm-related materials, such as gunpowder, gun-cleaning oil, shotgun wads, and lead wipes
- Fragments of packaging materials, such as paper, Styrofoam, plastic, and tape
- Any other materials in common use in the region served by the laboratory

Each sample of a reference material should be deposited onto several types of fabric, including cotton used in shirting or bed sheets, cotton denim, polyester used in work clothes, and, if frequently encountered in casework, wool. Materials will deposit differently onto each of these substrates. It is useful to have several examples from each common class of materials; for example, different types of facial cosmetics. It is also useful to prepare exemplars by direct touch contact, both light and heavy, smearing, and, in the case of a liquid or liquid suspension, dripping and spatter. Lastly, some of these exemplars should be touched to fresh pieces of fabric as examples of indirect transfer.

Reference collections of alteration and wear can be compiled from clothing items donated by persons who know their history, whether it be moth-eaten wool; pants singed at a campfire; blue jeans used while welding, painting, or working on cars; thinning of fabric from long use; or simply wear from repeated use. Simulations of wear are difficult, but garments can be purchased for use in certain activities so that specific types of alteration can be evaluated, and can be worn by someone willing to keep a log of the use history. Examples of mold and fungal growth on cloth items should be examined as they are encountered; the reference collection can include photographs of the actual growth.

Appendix 4

Sorting Tools for Stains and Deposits

What does it look like? Sorting tools for type of deposit

What does it remind you of?
Was the material deposited as a solid or liquid? As something in-between such as an emulsion (immiscible liquid in a liquid carrier) or suspension (undissolved solid in a liquid carrier)? As a mist or aerosol? Is it homogeneous or heterogeneous?

What is it? Sorting tools for type of material

Natural or synthetic: In general, natural structures exhibit more irregularity than synthetic processes or synthetic materials. For example, silk and polyester fibers are both extruded, the former from the body of the silkworm and the latter through a metal spinneret, and are then drawn, around a spool for polyester fibers and around the cocoon for silk. Silk, but not polyester fibers, exhibits subtle irregularly spaced swellings and constrictions along the shaft. Polyester fibers are usually symmetrical, with any planned variations regularly spaced. In comparing threads securing a button sewn on by hand with threads of a button machine-sewn at the factory, the hand-sewn button threads usually enter the fabric more randomly, whereas the threads of the machine-sewn button almost all pass through the same spots in the fabric.
Plant or animal: Plant cells, including wood structure, can usually be observed, if not further identified, using a stereomicroscope. Animal cells usually cannot be resolved using a stereomicroscope, although there may be hints in the stringiness of striated muscle tissue in meat. Animal tissue is often glistening and a bit slimy or stringy. Exceptions are bone, keratin (nails, hooves, and hair), and helical structures found in meat. Bone is porous, with chambers of varying size visible under the stereomicroscope. Processed plant material (i.e., food) can also be slimy or stringy, but usually in conjunction with visible plant cells or spheroid aggregates of starch grains. Occasionally, the structure of a synthetic composite can mimic cell structure when observed at low magnifications but can usually be distinguished at higher magnifications. The surrounding material should provide more clues. However, at this stage the focus is sorting, not identification, and providing descriptions of materials well enough to recognize potential sources on other items of evidence.
Metals, ceramics, plastics, minerals: Under the stereomicroscope, these materials look much like the corresponding larger objects that we see in daily life.
What processes formed the material, whether natural or artificial? (See Chapter 8 for a partial list.)

How did it get there? Sorting tools for manner of deposit

Was it deposited by contact, movement, projection, or a combination?
Did it originate from the inside or the outside of the garment?
What is the direction of smears or abrasions?
Does it overlap with other stains or deposits, or is it separate? If overlapping, are there portions that are separate?
Is it related to other deposits or to damage?
(Also see Chapter 5 for bloodstain pattern terminology.)

What happened to it? Sorting tools for alteration and wear

Wear: Effects of contact such as abrasion, scratches, cracks in the surface of crusts and films, flaking and other gradual loss of material. Note that when these processes occur in a single incident, tiny particulate residues produced at and loosely adhering to fracture margins indicate a relatively undisturbed (recent) deposit. These are lost with use and laundering.
Accumulation of material: Materials deposited over time, such as ground-in deposits, accumulations of loose debris, lint, variegated dust
Fading or discoloration: Alteration in color of stains that have been exposed to light and air over time
Damage: Is it fresh or older? Older damage is usually accompanied by evidence of wear or accumulated deposits in exposed damage surfaces

Sorting tools for examining samples received from another examiner

When a sample is received for further testing from the scientist who has examined the clothing, the sample should be described. Not only does this provide a record of what was received, but it also provides enough information so that subsampling can be documented and results of testing attributed to a known part of the deposit and to a component or location that has been described. Whether it was smeared on, stuck in, loose, originally liquid, or originally dust can be significant to an interpretation of the results. Similarly, it is important to know whether a sample was mixed in with, deposited on, or wadded up with other material. Information about any alteration, wear, or patterns should be noted. If a stain was visualized with an alternate light source and a sample cut out, the scientist examining the sample should check to see if any deposits are visible. Although it may seem to be repeating work already done, that may not be the case, because the original examiner may have had a broader focus and not have noticed as much detail. To review:

Description of stain, deposit, or damage
How the sample was deposited (smeared on, stuck in, loose, originally liquid, or originally dust)
Whether mixed in, deposited on, or wadded up with other material
Whether a stain is visible or detected with chemical tests or alternate light sources
Whether a stain or deposit is part of a pattern
Whether the stains, deposits, or damage appear recent with respect to last use

Index

C

D

E

F

G

H

M

N

O

P

Q

R

S

T

U

V

W

Y

Z